D1266483

sustainable*architecture*

low*tech*houses

Author: Arian Mostaedi
Publishers: Carles Broto & Josep Mª Minguet

Editorial team:
Production & Graphic Design: Francisco Orduña
Editorial Coordinator: Jacobo Krauel
Architectural Adviser: Pilar Chueca
Layout Design: Joan Fontbernat
Text: Jacobo Krauel unless the architects

© All languages (except Spanish language)
Carles Broto i Comerma
Ausias Marc 20, 4-2. 08010 Barcelona. Spain
Tel.: +34-93 301 21 99 · Fax: +34-93 302 67 97
info@linksbooks.net · www.linksbooks.net

ISBN: 84-89861-78-1
D.L.: B-13888-2002
Edition 2003

Printed in Spain

No part of this publication may be reproduced, stored inretrieval system
or transmited in any form or means, electronic, mechanical, photocopy-
ing, recording or otherwise, without the prior written permission of the
owner of the Copyright.

sustainable architecture

low tech houses

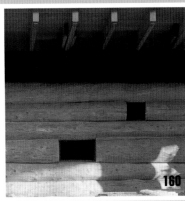

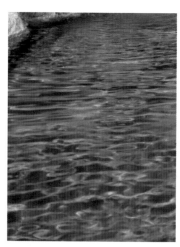

Introduction

For years there has been a general lack of ecological concern in construction. As the art that creates spaces for living, thus establishing the relation of man with his environment, architecture should be one of the disciplines in which the ecological spirit reaches its maximum expression. However, the majority of architects totally ignore the problems affecting the health of our planet, and continue to pollute in the name of architecture.

Though construction can still largely be seen as a stain on the landscape, several architects with ecological awareness have now stood aside from this tendency and aim to show the possibilities offered by sustainable architecture. The use of ecological or recycled materials, self-sufficient energy systems and systems for achieving a good temperature without heating can make a building something more than an artificial volume in the landscape and turn it into a space in which the built and nature respect each other.

This book offers a sample of some of the works that, through their originality, ingenuity and creativity, have found a method of sustainable design without the need for sophisticated cutting-edge technologies. In these designs—mostly dwellings—an interest in nature goes hand in hand with the design, providing beautiful spaces for living in harmony with the environment. They include houses made of adobe, rammed earth, bales of straw, wood, bamboo and recycled materials such as tyres and paper. In some cases they may be understood as a rediscovery of traditional methods that have been partly forgotten since the discovery of concrete. This, together with technological innovation, has made it possible to establish the keys to sustainability as it is understood by the architects that appear in this volume. This wide and varied selection includes works by famous architects in this field such as William McDonough, Shigeru Ban and Simón Vélez, which show that ethics in architecture is essential for the consolidation of this awareness if one wishes to avoid creating an uninhabitable world.

Hubbell & Hubbell Artists & Architects

deRenouard Residence

Jamul, California. USA

PHOTOGRAPHS: JOHN DURANT
DREW HUBBELL

Five years ago the owner of this residence found a piece of land in Rancho Jamul States he though was good deal, so he bought it on the spot. Although he was thinking of it as an investment purchase, when his wife saw that the property overlooked her favorite spot in all of San Diego County—Mt Kuchumaa—she knew immediately that she wanted to build a home there. They wanted a good relationship of indoors to outdoors, they wanted to celebrate the special view, they wanted energy efficiency and natural materials, and they wanted it all done in style. She was developing a keen interest in feng shui, as well, so it was important to her for energy to flow harmoniously through the house. While we are not feng shui experts, we have studied it enough to realize that its basic tenets align very closely with good design principles. "Good feng shui" can be achieved through sensitivity to a structure's environment and aesthetics.

By the time they came to us, the deRenouard had heard not only of James Hubbell's unique designs, but also of Hubbell & Hubbell's works using environmentally friendly alternative construction materials and methods. They were open to the idea of building with straw bale, a method he had used to very different aesthetic effect on a nearby project. Superinsulated straw bale walls turn a building into a kind of thermos, holding cool air in during the summer and the heat during the winter. California Title 24 energy calculations show that the house exceeds standard compliance by 31.2%.

The deRenouards were ahead of their time in anticipating the current energy crisis. Creating an energy-efficient and ecologically friendly home eventually included partially embedding the north edge of the building into the hillside, straw bale construction, solar hot water panels, high ceilings and operable transom windows, large winged roof overhangs on south-facing walls, slate flooring over concrete slab, plaster walls, blown-ill cellulose insulation, "energy star" Viking appliances, and native, low maintenance landscaping. Our first order of design was to situate the house on the long narrow lot in such a way that it would take advantage of both the views and the breezes, as well as passive solar effects. The deRenouards wanted a compact three bedroom home with a central living area separating master and guest bedrooms, as well as pool, and office / cabana (with steam room) by that pool, and a double garage with attached guest room.

Our solution to the challenge of the site was a series of interlocking blocks each set at an angle on the lot. Looked at from the air, it has the appearance of a double chevron, or an open zipper.

The façade of the house is a series of angled planes that are Modernist in a style drawn from Erik's Danish heritage. The flat lines of the parapet taper in seven places to a slight dip in the middle from which a line of blue tile pours down, visually drawing the intense blue of the Jamul sky down toward the earth like a stream of water. The pool is placed to frame views from the master bedroom and great room to the surrounding mountains. The deRenouards chose a beautiful slate tile from India for all the floors and the fireplace wall, and custom tile detailing in the James Hubbell style for the pool, the shower, and the hallway. The patio spaces are an integral part of the living space on both sides, acting as outdoor rooms between the garage / apartment, main house, and office / cabana.

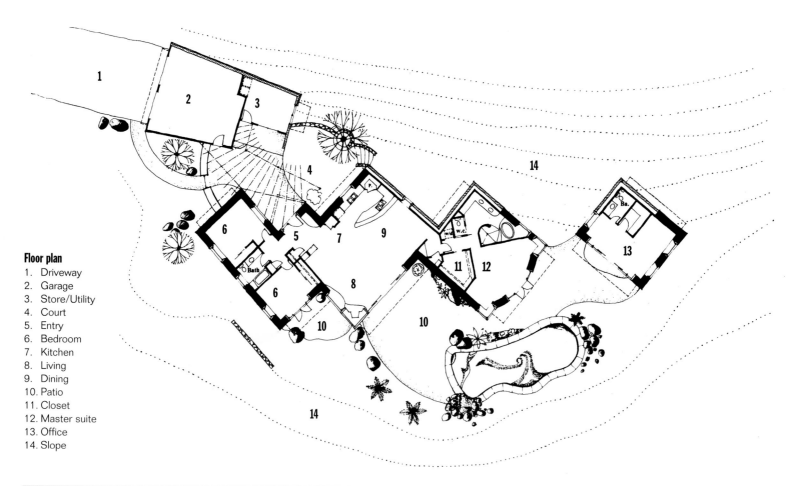

Floor plan

1. Driveway
2. Garage
3. Store/Utility
4. Court
5. Entry
6. Bedroom
7. Kitchen
8. Living
9. Dining
10. Patio
11. Closet
12. Master suite
13. Office
14. Slope

The construction with bales of straw is totally ecological and allows the dwelling to become a kind of thermos that can keep the air cool in summer and warm in winter. The construction process is simple and plaster or stucco can be used for a finish.

Forsythe + MacAllen Design Associates

Mitchell / DeCairo House and Art Gallery

Galiano Island, British Columbia. Canada

PHOTOGRAPHS: Forsythe + MacAllen Design Associates

The architects built this house, in British Coluimbia's Gulf Islands, for an artist and a teacher. The long narrow house, which also acts as an art gallery, consists of a series of rooms arranged along a hallway.

Following the space of a natural stone plinth, the house is slightly kinked. The shape is perceived in the gallery / hallway as a transition from the public to the private spaces of the house. The exposed structure and cladding for this house were selectively horse-logged and milled on site. The yellow cedar floor was machined from wood that the architects salvaged from the deck of a local tennis court. Other elements such as doors and some of the built-in furniture were also built from salvaged lumber.

The house's form was partially influenced by the old timber utility structures typical of abandoned mining towns in British Colombia. What is interesting about these buildings is the way they sag and flex as they are age and settle into their foundations. As the architects built the house they further fine-tuned the slight shifts from particular views and qualities of light.

In the summer, when the house also operates as a gallery, its large doors can be opened up so that the expansive decks surrounding the building contribute to its living space. This helps circulate and distribute large crowds of people. In doing this a sense of domestic intimacy is maintained.

Also built on this site are two utility pavilions. Aside from practical storage these little buildings bring the owners two parts of their property that they would otherwise rarely visit. These structures where designed so that they could be carefully located amongst the trees of an old growth forest without disturbing their ancient roots. One building is a tool shed. Its cladding, which is charred black and oiled reflects the colour of the forest. The other building, a woodshed, was designed to store and season firewood that is selectively harvested off the property. The firewood, used to heat much of the house, is sheltered by a wide overhanging roof manufactured by laminating thin sheets of bent plywood into a single shell.

The form of this dwelling is influenced by the old wooden structures typical of the abandoned mining cities of the zone.
All the wood used was transported without using polluting transport and was treated on site.

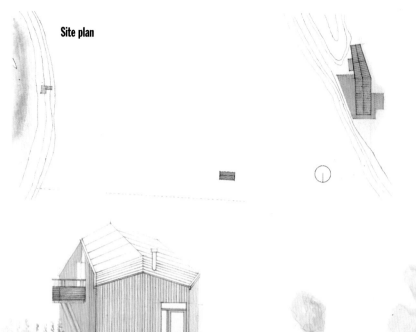

Cross-section

North elevation

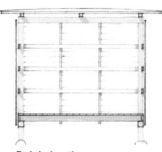

Tool shed section

Tool shed elevation

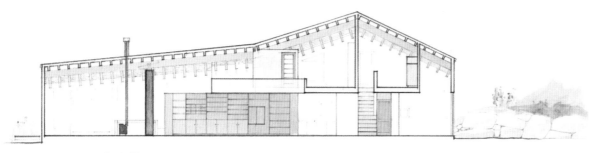

Long gallery section looking east

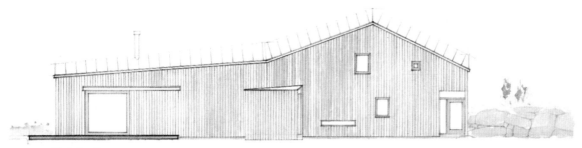

West elevation

1. Entry
2. Living
3. Dining
4. Kitchen
5. Library/Computer gallery
6. Dark room/wine
7. Multi-purpose
8. Technical
9. Laundry
10. Bath
11. Guest bedroom
12. Main bedroom
13. Clothes storage
14. Stamp collection
15. Bath
16. Sewing loft

First floor plan

Second floor plan

In the design of the construction system it was decided to combine the advantages of a solid construction (heat storage, acoustic protection) with the advantages of light walls (no desiccation, good thermal insulation, rapid regulation of the temperature of rooms…).
All the wood used is autochthonous and the other materials are also natural and recyclable.

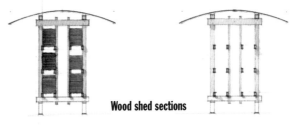

Wood shed sections

Jones Studio, Inc
Johnson-Jones Residence
Phoenix, Arizona. USA

PHOTOGRAPHS:TIMOTHY HURSLEY
SALLY SCHOOLMASTER

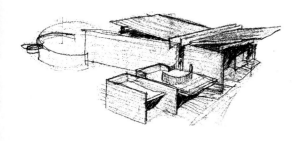

No one wanted the 1 acre mountain preserve land with a natural 16,000 acre backyard because, everyone knew that the proximity of the huge, neighbourhood chlorinating tank was unacceptable.

Just a few simple moves created a rammed earth form that integrated the beautiful mountain views due north, with a softly daylit family gathering space. And each cylindrically composed outdoor room made friends with the Mother cylinder.

The house fits between the existing water tank, the mountain preserve and the pink stucco, red tile roof neighbouring homes. It's cylindrical site walls and north-south facing fenestration strategically capture mountain views while masking unwanted sights of the water tank and adjacent neighbourhood. The rotated concrete, bubble block walls used in the front and back of the house afford privacy without sacrificing air circulation.

Known by the community as "The Dirt House", the residence reintroduces rammed earth, a 1000 year old building technology, in a contemporary, resource-efficient way. To create the rammed earth wall, scoops of barely-moistened dirt from the site, mixed with 3% Portland cement, are loaded into plywood forms in lifts of eight inches. A hand-held pneumatic tamper compacts the dirt into a rock-hard six-inch layer. This process is repeated until the desired wall height is achieved. The forms are removed, resulting in a very solid, energy-efficient wall system. The rammed walls of this residence are 2 feet thick and 18 feet tall. The heat will conduct through compacted dirt at a rate of 1 inch per hour, so, the interior surface maintains a constant room temperature throughout the year.

Most of the construction materials used in this home have significant recycled content and/or are resource-efficient materials. A few notable examples are: plywood glued together with toxin-free adhesives, low to no VOC paints (actually, there is almost no paint on the project) high performance glazing, engineered composite framing lumber, rusted steel wall cladding, and concrete block manufactured with fly ash.

Extensive overhangs, a rusted steel louver, and careful solar orientation ensure complete natural day lighting with responsible sun control.

Landscaping in the yard is naturally irrigated by a rainwater-harvesting system. Rain water is directed gently from the roof to the desert floor by a stainless steel chain-link hung from rusted, black iron steel plates. The roof is pitched toward the large scupper projecting from the stair tower. Rainwater clings to "rain-chains" and is deflected into an 18-foot diameter holding area by large concrete sphere splash blocks. When the detention area fills, the water overflows into the landscaping.

Every habitable space is bathed in natural, filtered daylight, substantially decreasing the need for electric lighting during the daytime hours.

East elevation

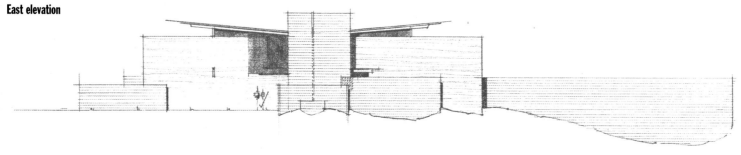

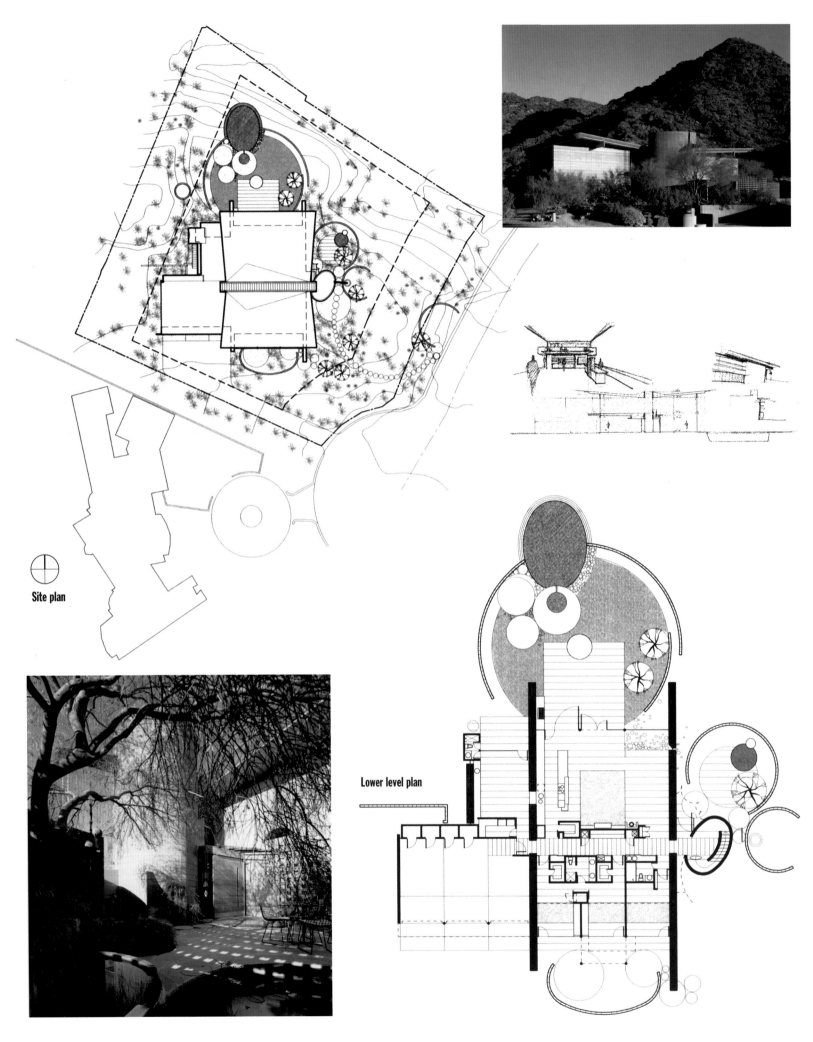

Site plan

Lower level plan

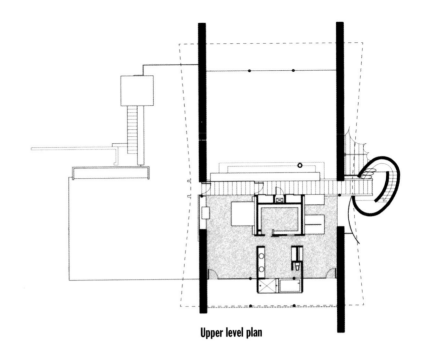

Upper level plan

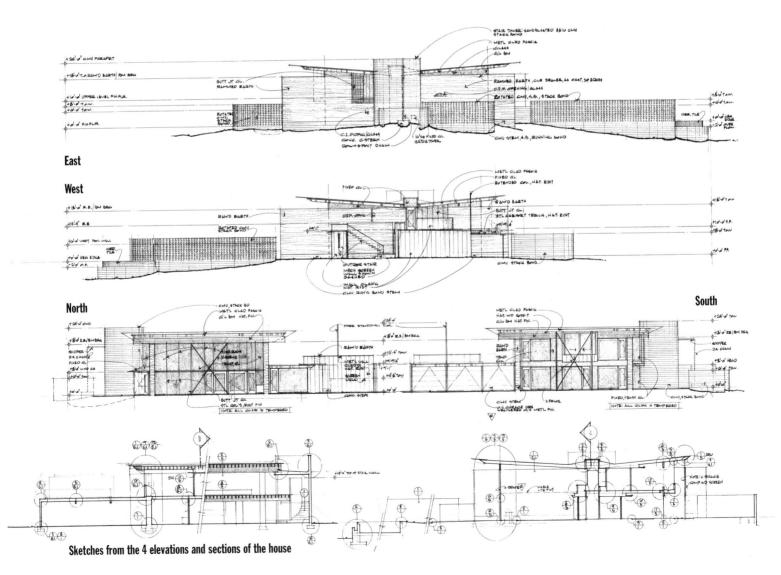

East

West

North   **South**

Sketches from the 4 elevations and sections of the house

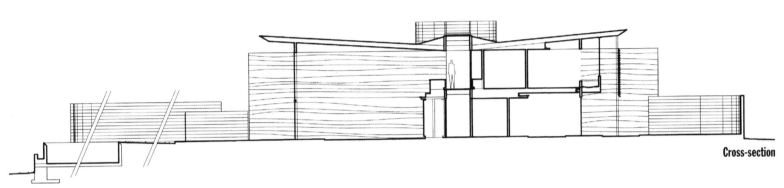

Cross-section

The main space of the dwelling combines all the public zones of the dwelling, including the kitchen and the dining room.
The transversal bracing system, right, crosses the large glazed window providing stability, support for the lighting, and allowing good views of the mountains.

Flemming Skude

Con Cave

Hummingen, Lolland. Denmark

PHOTOGRAPHS: FLEMMING SKUDE

Somehow this grounded house represents quite the opposite of Glenn Murcutt's credo: "Touch this earth lightly" although it could also be seen as a current exponent for a the building tradition in Denmark. Nature-integrated and aerodynamic dwellings can be traced thousands of years back to the Iron Age. At that time dwellings used to be sunk about a metre into the terrain and covered by a thatched roof. Even built on hills in the landscape problems of flooding and water in heavy rain periods could occur.

(In Iceland, the Faroe Islands and in Greenland this housetype has survived up to our days with stones and turf on the outside and wooden panels on the inside).

During the Viking era (ca.750-1050 AD) solitary buildings often had a ship-like plan since carpenters from shipmaking knew ou well that a rectangular form acted badly against waves of the sea and forces of the wind. Because of this latter fact the CON CAVE house has its maximum width and height along its middle and is crowned by a skylight at the top of the roof where its predecessors used to have an opening for letting out smoke from the fireplace. In fact the details of the ceiling in CON CAVE look very shiplike from the inside.

From the outside the house recalls dolmens or cromlechs of the Bronze Era, also since the main entrance is flanked by stones in the way used for graveyards in those days. This means that from the outside this house is expected to be very cold and dark on the inside –quite the opposite of reality. In fact CON CAVE should be understood as a wooden coffin placed above the surrounding terrain and protected sidewards by a wall of concrete blocks preventing the covering soil sides from pressing the facades towards the house. Oriented strictly on an east-west axis, the earth coverings make 3 natural terraces to the south, the east and the west. Technically the inside wood construction has a normal insulation of 10 centimetres and is ventilated on all sides as well as in the roofing. The outside of the wood coffin has a strong Platon membrane (normally used on external basement walls) protecting against humidity and moisture. On the top of this membrane there are two layers of turf about 12 cm thick.

By covering a "normal" wood construction by earthern slopes and grass turf, the final outcome is an overall aerodynamic form and a mayor improvement in natural temperature control since CON CAVE is never under +5° C during the winter. The house also has a pleasant cooler average temperature than the outside during the summers—without any artificial ventilation devices. The owner of the house in fact never wanted a hearth in the first place but has installed a few electrical stoves—just in case… The outcome of this simple house with small facades covered with copper at top—ike the cladding of the skylight—is a minimum of maintenance, a cosy dwelling and almost no heating expenses. The 'horns' over the gables to the east and the west are intended for hanging up sun-protecting sails, but so far have never been put to work.

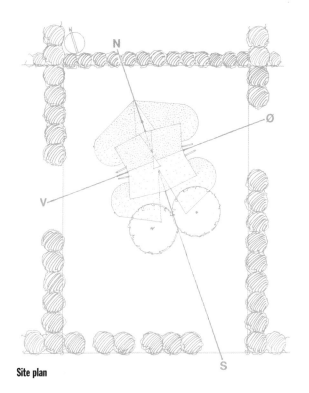

Site plan

Floor plan

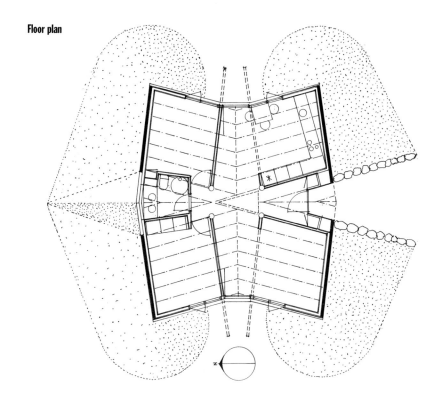

Following ancient techniques of the traditional architecture of northern Europe, the dwelling appears semi-buried in the land. This natural integration allows the wild grass to completely cover the roof structure, leaving exposed only the access zones.

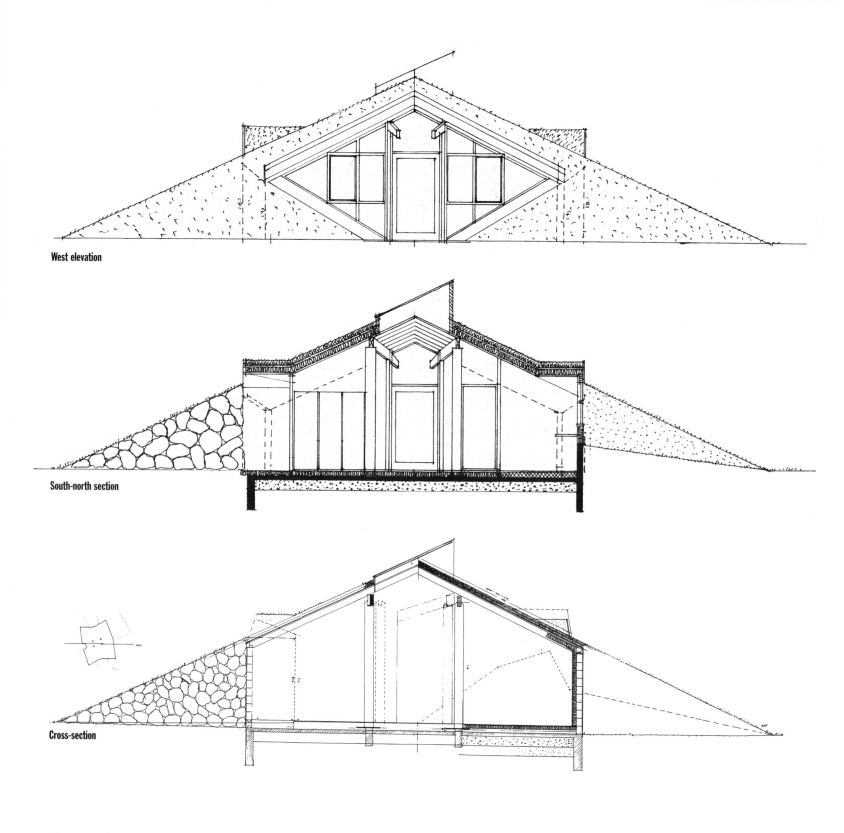

West elevation

South-north section

Cross-section

The design of the house is aerodynamic to offer better protection against the weather. To enhance the entrance of natural light, a triangular skylight stands out through the vegetation cover.

Simone Swan
Swan House / Parr House
Presidio, Texas. USA

FOTOGRAPHS: WAYNE BINGHAM
RONALD RAEL (PARR HOUSE)

Simone Swan is an architect specialised in the construction of adobe dwellings that are particularly suitable for withstanding the harsh climate of desert areas. This type of building has a firm structure with vaults and domes that recall traditional Egyptian dwellings. In fact, Swan studied under Hassan Fathy (1900-1989), an Egyptian architect who thought that buildings should be treated as a dynamic extension of public and private life. The work is thus generally done in teams, together with the owners and volunteers who want to learn this type of building technique.

Her experimentation over the years has shown her that buildings made with adobe bricks are more economical and ecological, and provide the best conservation of the indoor temperature. The roofs and domes are built without using any wood, thus helping to avoid deforestation.

The H-shaped floor plan allows two courts to be created, one on the east side and one on the west side. Different habitable spaces are thus created both on the exterior and on the interior, making optimum use of sunlight and shade according to the time of day and year.

The aim is not only to build a dwelling, but also to create a space that is as pleasant due to its external image as due to its interior comfort, and that this should occur in such a way that it does not alter nature during the building process or waste energy unnecessarily. Therefore, these dwellings are ready for the fitting of solar energy accumulation systems.

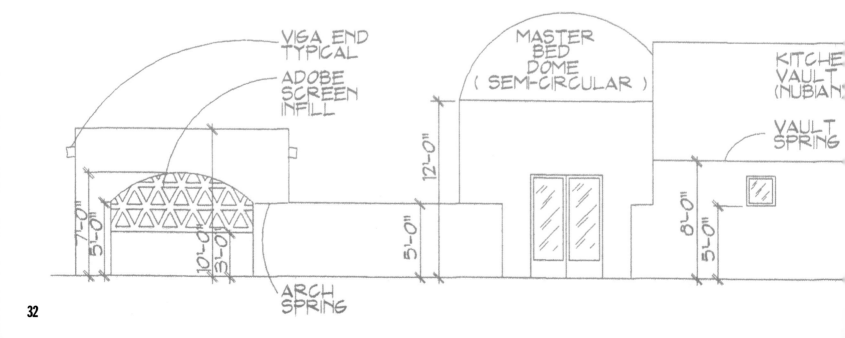

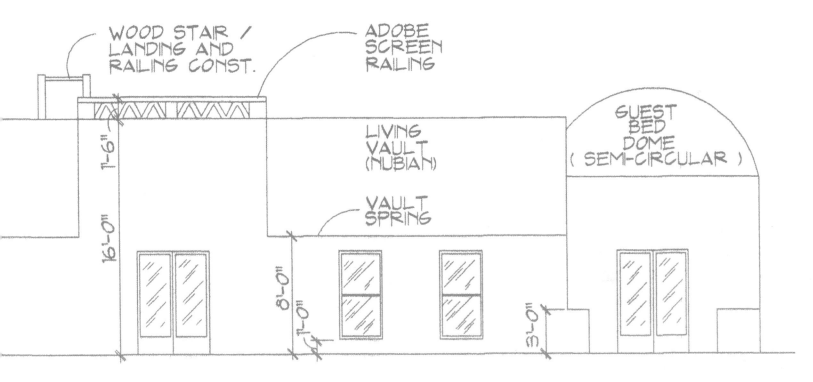

WOOD STAIR /
LANDING AND
RAILING CONST.

ADOBE
SCREEN
RAILING

GUEST
BED
DOME
(SEMI-CIRCULAR)

LIVING
VAULT
(NUBIAN)

VAULT
SPRING

1'-6"

16'-0"

8'-0"

1'-0"

3'-0"

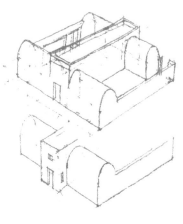

This page and the previous page show details of the dwelling of Simone Swan.
In the interior, the large spaces maintain a pleasant constant temperature thanks to the
use of adobe. The house can thus withstand the sharp differences in temperature that
occur between day and night in desert areas without using heating.

The adobe bricks are laid using a completely manual technique. The vaults were built without moulds so as to avoid the use of wood or other materials.

The H-shaped floor plan creates two large courts that are protected by two adobe walls. One of these walls is designed to allow gentle sunlight to filter through.

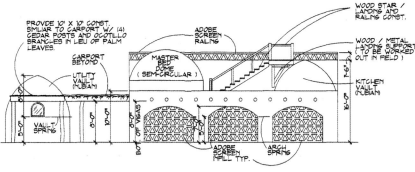

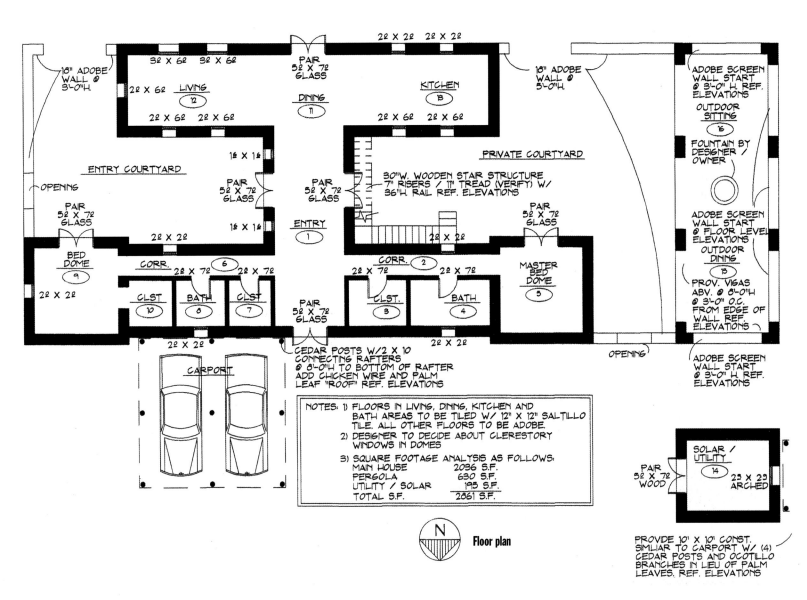

NOTES: 1) FLOORS IN LIVING, DINING, KITCHEN AND BATH AREAS TO BE TILED W/ 12" X 12" SALTILLO TILE. ALL OTHER FLOORS TO BE ADOBE.
2) DESIGNER TO DECIDE ABOUT CLERESTORY WINDOWS IN DOMES
3) SQUARE FOOTAGE ANALYSIS AS FOLLOWS:
MAIN HOUSE   2036 S.F.
PERGOLA   630 S.F.
UTILITY / SOLAR   195 S.F.
TOTAL S.F.   2861 S.F.

Floor plan

Love-Architecture & Urbanism

Softech-home; Solar house

Gamlitz, Styria. Austria

PHOTOGRAPHS: ANDREAS BALLOON
LOVE-ARCHITECTURE & URBANISM

This is the prototype of a system house destinated for series production. Softech, a specialist in low-energy construction, launched the idea of developing a system house combining highest aesthetic standards with up-to-date low-energy technique.

The house was conceived as a "summer-winter-house": The living room (=winter-living room) corresponds in size with the terrace (=summer-living room); both are connected by a large sliding glass-wall. Bathroom and kitchen can be used equally from the summer—and winter—living room. This concept enlargens the house beyond its built surface.

We tried to extend the usefulness of the house in summer by creating a terrace protected from wind and sight in front of the living room and connecting the two by a large sliding and revolving door. We also optimised the room sur-

faces in regard of their use and minimised corridors and service spaces. Developing the construction system, the following requirements have been given priority: 1) The combination of the advantages of massive construction (heat storage, acoustic protection) with the advantages of lightweight walls (no desiccating time, good heat insulation, quick regulation of room temperature…). 2) The employment of natural, regenerating materials, using indigenous wood.

The outer walls were constructed as follows:

Exterior surface: 1) Non-treated, glued larch boards. 2) Cellulose insulation. 3) Natural gypsum fibreboards, partially clay plaster.

The estimated time for building this dwelling was 6 to 8 months, a very short time bearing in mind the quality of the finishes and its floor space of 156 m^2.

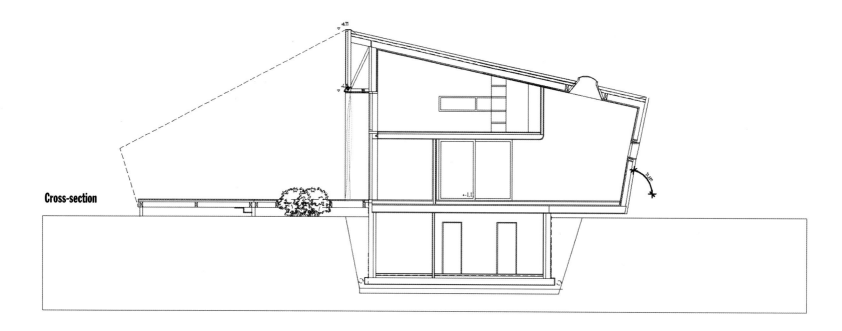

Cross-section

This solar house is the largest of a series of dwellings made in conjunction with Softech, a company specialised in research and development of renewable energy systems applied to construction. The spaces were configured so that the dwelling could be used equally in summer and in winter.

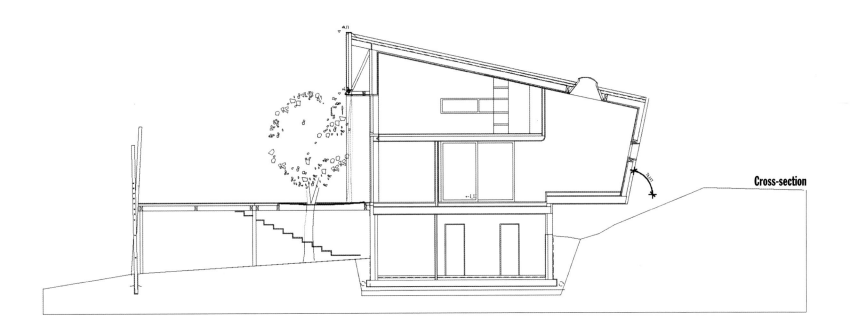

Cross-section

In the design of the construction system it was decided to combine the advantages of a solid construction (heat storage, acoustic protection) with the advantages of light walls (no desiccating, good thermal insulation, rapid regulation of the temperature of rooms…). All wood used is autochthonous, and the other materials are also natural and recyclable.

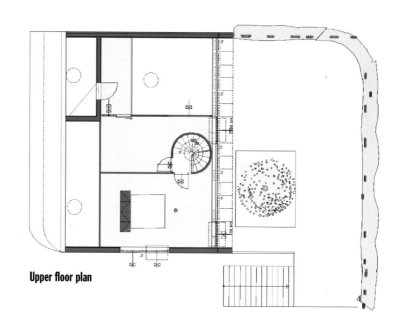

Upper floor plan

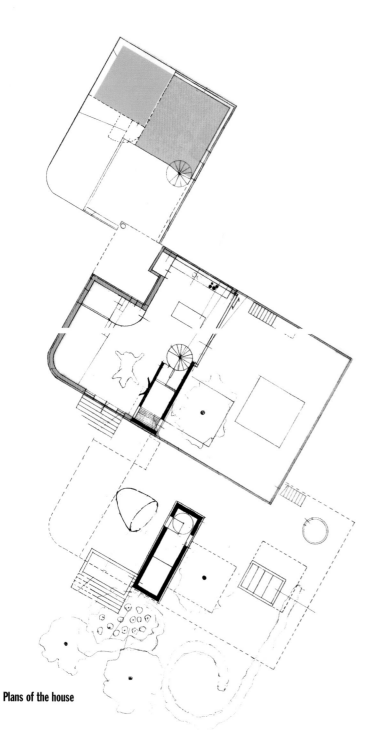

Plans of the house

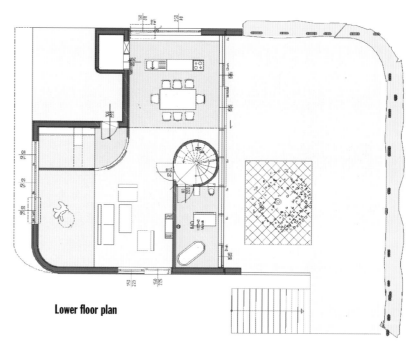

Lower floor plan

Leventhal Schlosser Architects
MacIver Residence
Mendocino, California. USA

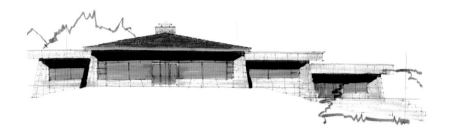

FOTOGRAPHS: DOUGLAS HILL

This dwelling of 300 m², located in a privileged area of the Californian coast with direct views of the Pacific Ocean, was designed with the idea that its appearance should create a link with the surrounding landscape. To achieve this, rammed earth was used to build the outer walls. This material has been widely used in the past due to its low cost and ease of handling, and the advantages of efficient thermal control of the interior. This scheme is a perfect example of the possibilities offered by this material, which was forgotten for generations due to the arrival of new mass-produced materials. In addition to its high quality and strength, rammed earth is completely ecological and safe, because it is completely natural.

The client brief consisted in developing a sequence of spaces that were comfortable, safe and exciting. Intended for a couple, the dwelling opens onto the exterior through large windows that fill it with daylight, thus helping to minimise energy use.

In addition to the habitual spaces, two studios and guest zones were created. An interesting feature is the main bathroom with a circular shower located under a domed skylight.

The form adopted by the dwelling is determined by the topography of the location and its orientation. The house is sunny and well-lit, avoiding great use of energy for heating and lighting.

Elevations

Floor plan

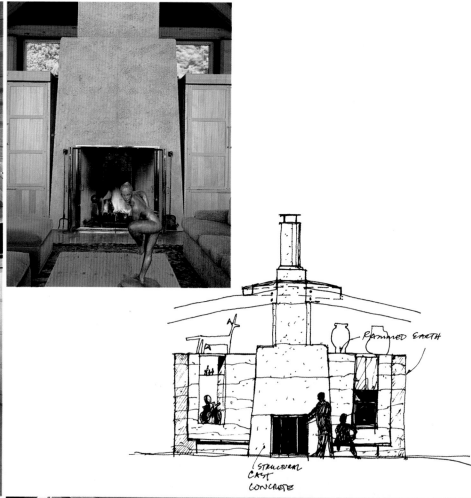

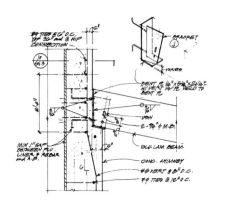

Construction detail of the chimney

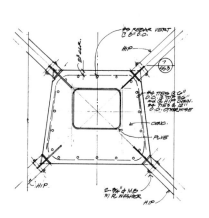

The interior spaces have privileged views of the Pacific Ocean thanks to the large windows and glazing along the facades.

In one of the studies it was decided to include a skylight in order to further increase the use of daylight.

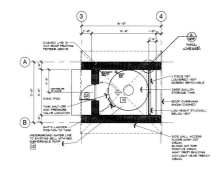

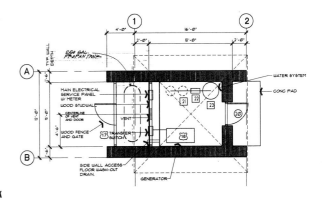

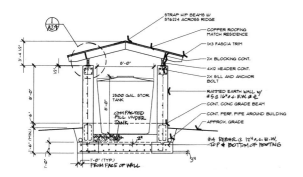

Construction detail of the water tank

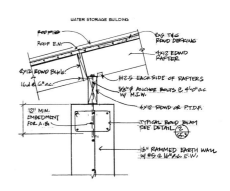

The placing of a dome-shaped skylight over the shower in the bathroom provides ample natural lighting and tempers the separation between interior and exterior.

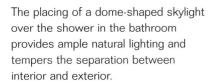

Gregory Burgess Pty Ltd Architects
Burraworrin Residence Flinders
Flinders, Victoria. Australia

PHOTOGRAPHS: TREVOR MEIN
GREG BURGESS

Located on the coast of the peninsula of Mornington, this dwelling designed for four families creates an awareness of the continually changing relations between daily activities and the natural profundity of the landscape, the seas and the sky. The scheme shows obvious analogies with marine aesthetics, so the dwelling appears to passers-by as a stranded boat awaiting a new journey. The accentuated and complex parabolic curves also suggest that Gregory Burgess was seeking organic forms in an attempt to link culture with nature.

In the construction of this dwelling, rustic wood trunks cut radially and limestone from the local coast were used to resist the harsh marine environment. Another reason for using radially sawn wood is the range of environmental advantages provided by this type of cut. The radial system can be extended to all plantations of trees farmed for wood, thus maximising production. Fewer trees are needed to obtain the same amount of useful wood, and this also causes a lesser environmental impact on the soil, the aquifers and the fauna. It also helps to reduce the need for other types of materials that may have negative consequences for the environment.

In the interior of the dwelling, the exposed limestone is combined skilfully with wood, creating a modern and welcoming environment. The views offered by the location, framed carefully by the protruding windows, are surprising in their beauty and offer a spectacular panorama of the coastal landscape.

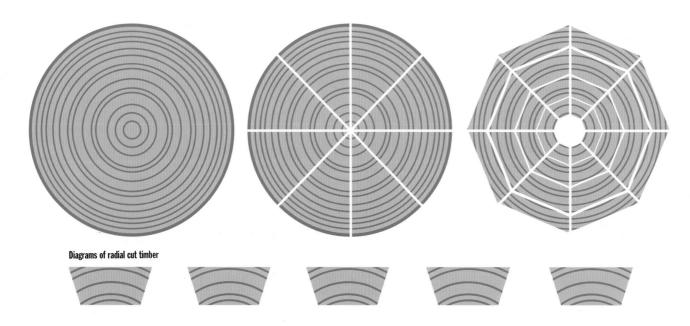

Diagrams of radial cut timber

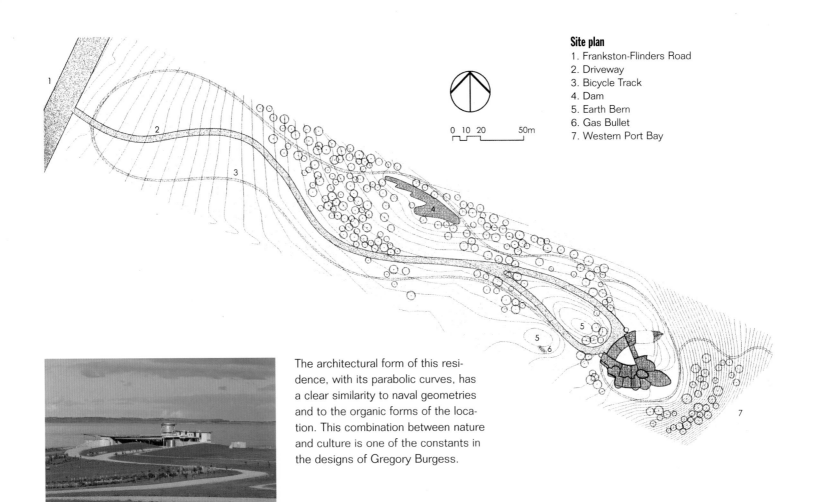

Site plan
1. Frankston-Flinders Road
2. Driveway
3. Bicycle Track
4. Dam
5. Earth Bern
6. Gas Bullet
7. Western Port Bay

0 10 20   50m

The architectural form of this residence, with its parabolic curves, has a clear similarity to naval geometries and to the organic forms of the location. This combination between nature and culture is one of the constants in the designs of Gregory Burgess.

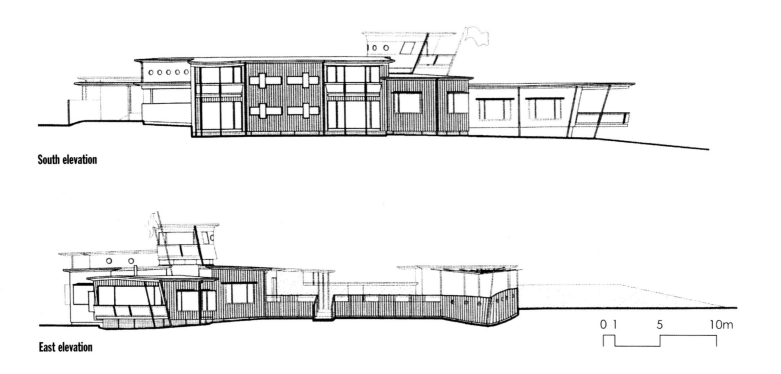

South elevation

East elevation

0 1   5   10m

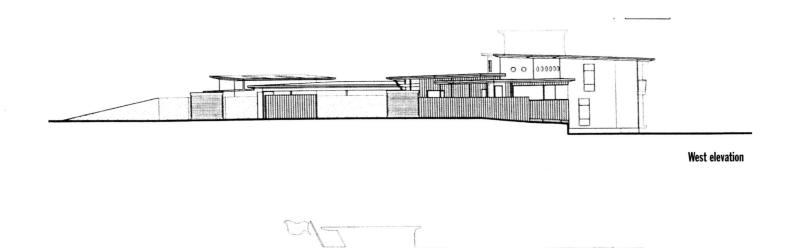

West elevation

North elevation

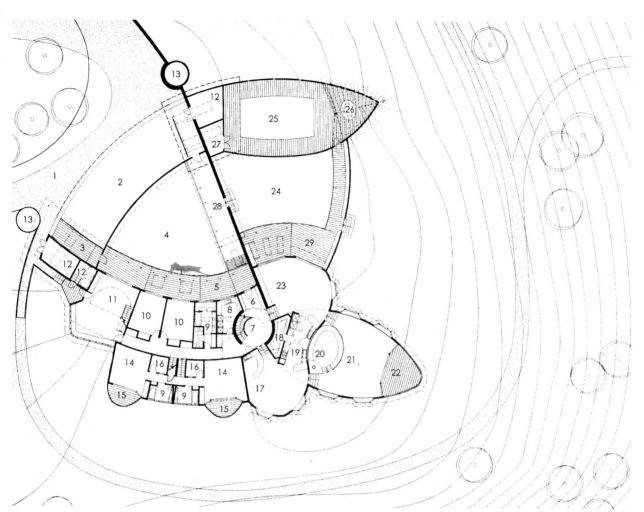

Ground floor plan

1. Driveway
2. Carport
3. Ramp
4. Courtyard
5. Verandah
6. Entry
7. Vestibule
8. Laundry
9. Bathroom
10. Bedroom
11. Rumpus
12. Store
13. Rain water tank
14. Master bedroom
15. Balcony
16. Alcove
17. Dining
18. Pantry
19. Kitchen
20. Firepit
21. Living
22. Deck
23. Breakfast
24. Terrace
25. Pool
26. Spa
27. Pool equipment
28. Walkway
29. BBQ

0 1   5

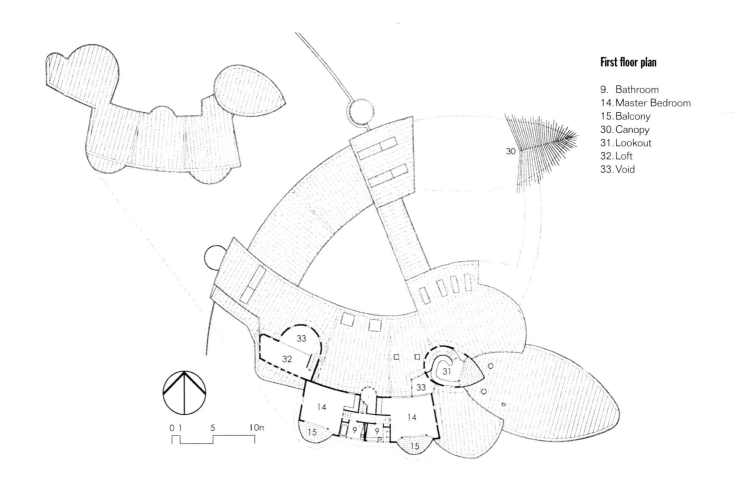

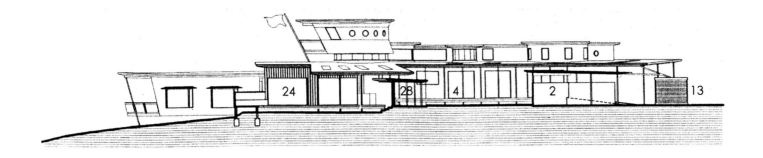

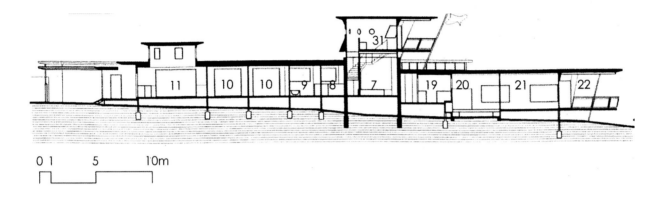

```
0 1     5      10m
```

Sections

2. Carport	11. Rumpus	21. Living
3. Ramp	12. Store	22. Deck
4. Courtyard	13. Rain water tank	23. Breakfast
5. Verandah	14. Master bedroom	24. Terrace
7. Vestibule	15. Balcony	25. Pool
8. Laundry	17. Dining	28. Walkway
9. Bathroom	19. Kitchen	29. BBQ
10. Bedroom	20. Firepit	31. Lookout
		32. Loft

All the wood has been treated following an innovative system of radial cutting that provides a series of ecological advantages. With this system the number of trees cut to produce a given amount of wood is considerably reduced.

The impact on the soil, the water and the fauna is also reduced, as is the need to use other types of materials that may have negative consequences for the environment.

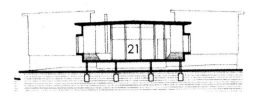

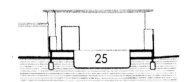

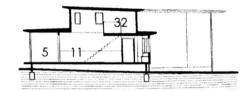

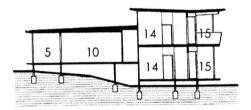

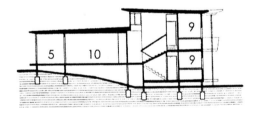

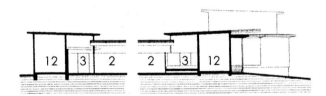

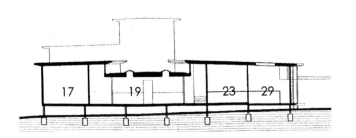

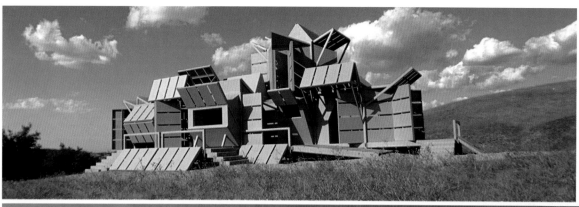

60

Michael Jantzen
M-House
Gorman, California. USA

Relocatable M-vironments are made of a wide variety of manipulatable components that can be connected in many different ways to a matrix of modular support frames. The frames can be assembled and disassembled in different ways to accommodate a wide range of changing needs. The M-house, which is made from the M-vironmens system, consists of a series of rectangular panels that are attached with hingeds to an open space frame grid of seven interlocking cubes. The panels are hinges to the cubes in either a horizontal or a vertical orientation. The hinges allow the panels to fold into, or out of the cube frames to perform various functions. Some of the panels are insulated and contain windows and doors. These panels can completely enclose spaces that are heated and cooled. Other un-insulated panels fold in or out over and around these open platforms. The platforms and the cube frames are supported by adjustable legs which are attached to load the bearing footpads. In many cases the support frames do not require a foundation, and they can be adjusted to accommodate terrain variations. All of the M-house components are interchangeable, and can be increased or decreased in number and size. The panels can be made in a curved configuration and from many different types of materials. The existing M-house was designed to function as a single private vacation retreat, or in multiple number and configurations, as a complete stand-alone high-tech resort complex.

The house can be designed to be self sufficient, powered by alternative energy sources such as the sun and the wind. With different sizes, shapes, materials, and panel types, the system can be used for exhibit structures, pavilions, play environments for kids, retail spaces, office modules, and many other commercial applications.

PHOTOGRAPHS: MICHAEL JANTZEN

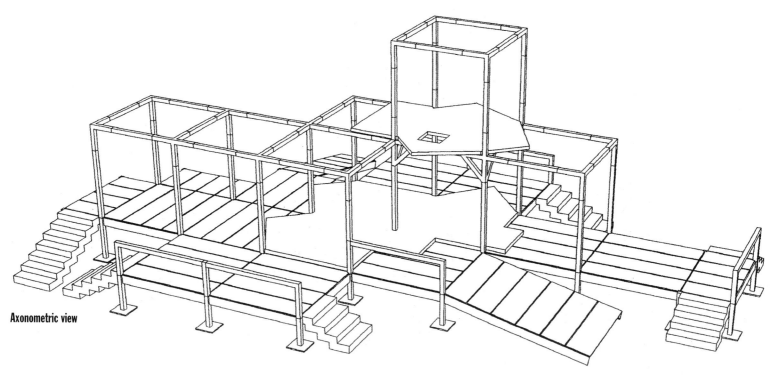

Axonometric view

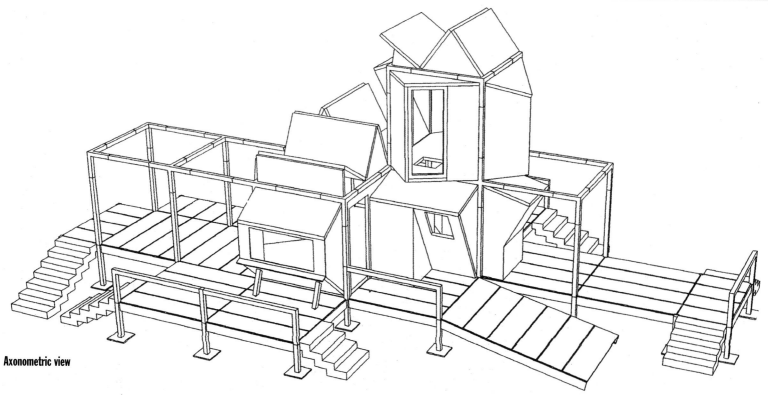

Axonometric view

Axonometric view

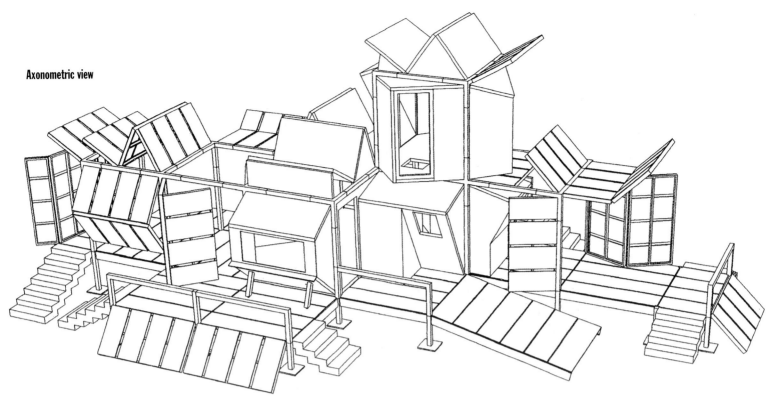

This dwelling is designed with a system of mobile panels that allow the morphology to be varied according to the needs of the occupants. Another advantage of this type of "light architecture" is that it can be adapted to almost any type of land because it has no foundations.

Floor plan

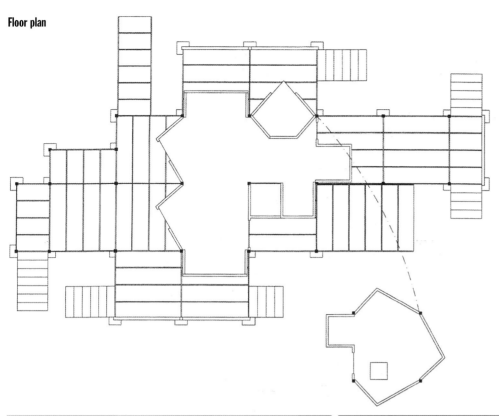

Thanks to the system of folding panels, the dwelling can be opened or closed to regulate the temperature of the interior. The spaces created also have optimum natural lighting thanks to the strategic arrangement of windows and skylights.

Arkin Tilt Architects
Caner/Beier Residence
Napa County, California. USA

PHOTOGRAPHS: ED CALDWELL PHOTOGRAPHY

Located on a steep, west-facing slope of the Lovall Valley, the residence nestles into a saddle in the land tucked behind a small knoll. The residence is segmented into three individual structures both to take advantage of the views and to reinforce the connection with the site: vistas are framed of the classic oak and grass landscape and of the spectacular views both west towards Sonoma and south towards San Francisco. Organized along the contour of the hill, the indoor rooms create places of privacy while forming more public outdoor rooms between them. Each room balances view and solar gain with careful window placement and shading devices.

The main spaces are of sprayed earth construction—providing both thermal moderation in its 18" thickness and a rich, patinaed finish—with a roof structure of recycled fir trusses and salvaged cypress decking. Rammed earth is used at the fireplace and loggia columns for drama. A corrugated metal roof and plywood and batten siding or cribbing on secondary spaces draw from the agrarian California vernacular, giving the buildings a more casual and modern feel.

Palter Residence
Sant Mateo, California. USA

PHOTOGRAPHS: ED CALDWELL PHOTOGRAPHY

The challenge in this project was to transform an ordinary 3 bedroom / 2 bath cramped "ranch -style" house with difficult orientation—the living room and main canyon views face north—into a sunny and liveable space for a couple with two small children. Keeping in mind a tight budget, the architects reorganized the plan creating separate zones for the master suite and the kid's rooms, including relocating the family room to the former master bedroom, which over-looks the south-facing garden terrace.

At the east end, minimal square footage was added around the former family room (added in the 60s) to create a master suite and tie the incongruous addition back into the house.

The walls and an obtrusive brick fireplace were removed between the kitchen, entry and living room, and the ceiling height was raised into the attic, using the existing structure and skip sheathing as the exposed ceiling. A cupola was added just south of the ridge, flooding the new "Great Room" with natural light and providing a high vent for natural ventilation. Large glazed barn sliders open the north wall up to a terraced patio facing the canyon.

Natural and recycled materials were used throughout—including madrone trunk columns, recycled glass countertops, and salvaged doors and light fixtures—both for lower environmental impact and for character.

Though the building does not have an ideal orientation, the main spaces are of sprayed earth construction—providing both thermal moderation in its 18" thickness an a rich patinaed finish— with a roof structure of recycled fir trusses and salvaged cypress decking.

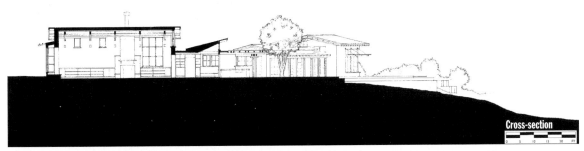

Cross-section

0 5 10 15 30 FT

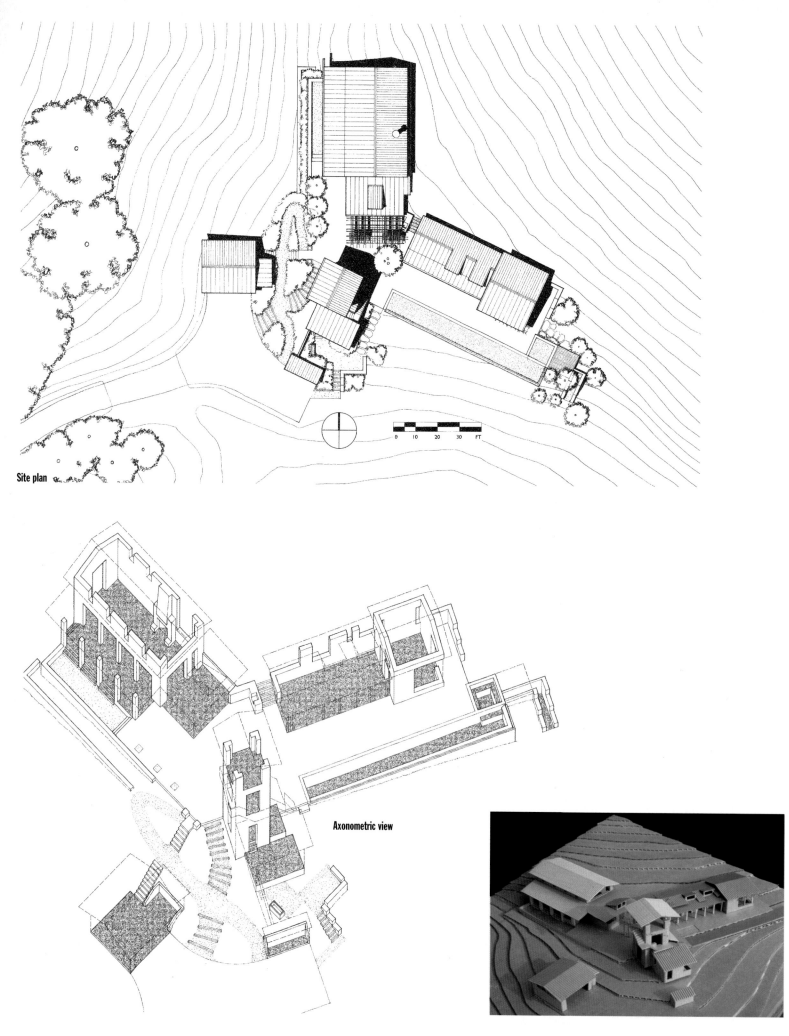

Site plan

Axonometric view

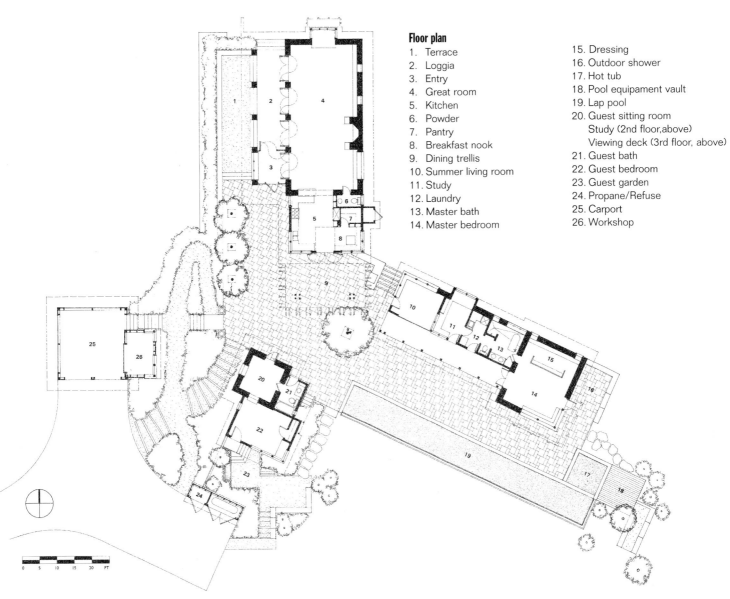

Floor plan

1. Terrace
2. Loggia
3. Entry
4. Great room
5. Kitchen
6. Powder
7. Pantry
8. Breakfast nook
9. Dining trellis
10. Summer living room
11. Study
12. Laundry
13. Master bath
14. Master bedroom
15. Dressing
16. Outdoor shower
17. Hot tub
18. Pool equipament vault
19. Lap pool
20. Guest sitting room
    Study (2nd floor, above)
    Viewing deck (3rd floor, above)
21. Guest bath
22. Guest bedroom
23. Guest garden
24. Propane/Refuse
25. Carport
26. Workshop

In the interior of the dwelling the architects have created large and well-lit spaces, such as the double height living room. Following an environmental approach, it was decided to use recycled Douglas fir beams and cypress boards from old roofs to build the roof frame.

Palter Residence

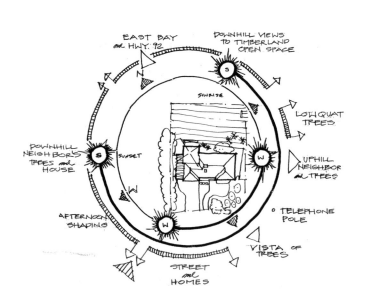

EAST BAY
on HWY. 92

DOWNHILL VIEWS
TO TIMBERLAND
OPEN SPACE

SUNRISE

LOQUAT
TREES

DOWNHILL
NEIGHBOR'S
TREES and
HOUSE

SUNSET

UPHILL
NEIGHBOR
on TREES

AFTERNOON
SHADING

TELEPHONE
POLE

VISTA of
TREES

STREET
and
HOMES

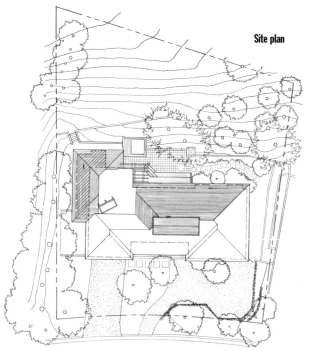

Site plan

Cross-section

0 2 4 6 8 10

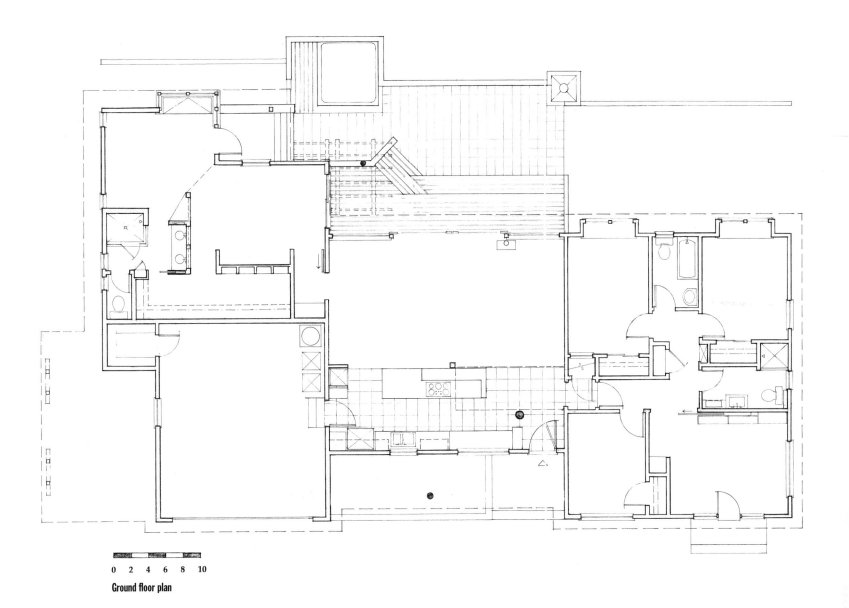

0 2 4 6 8 10

Ground floor plan

Jüngling & Hagmann
Atelier und Wohnhaus in Lüen
Lüen, Switzerland

PHOTOGRAPHS: SCHENK & CAMPELL

In the centre of Lüen, a former raft-making town, an old farm building was to be replaced by a new building. The external dimensions of the building were laid down by the planning regulations, which allowed a new building to be constructed only if it matched the dimensions of the original building.

The layout of the modest ground floor determined its new configuration and its new use as a photographic studio and holiday home. However, the architects did not imitate the traditional vernacular of the town, but took advantage of the legal restrictions to make a careful reinterpretation of the traditional building. The final result is a building that is perfectly neutral in its environment but far from sterile in its interpretation of the site and of the architectural and functional needs.

For the construction it was decided to used trunks of rustic timber, placed as an open cladding within a geometry of horizontal stripes. On the facade, the traditional local use of trunks to build sheds was newly interpreted.

Industrial birch panels were used to clad the interior walls. In the bathroom and the kitchen, which appear to be elements of the furniture, coloured wood panels were placed in accordance with the uniformity of the whole.

The whole house is perfectly insulated and the window panes are specially manufactured to provide protection against condensation.

Construction using local timber trunks is a common practice in this region of Austria. It integrates the new building in the landscape and evokes the customs and traditions of this mountain region. This relation between lifestyle and architecture is also in harmony with nature, and in the construction the use of materials that could be toxic or unecological was avoided.

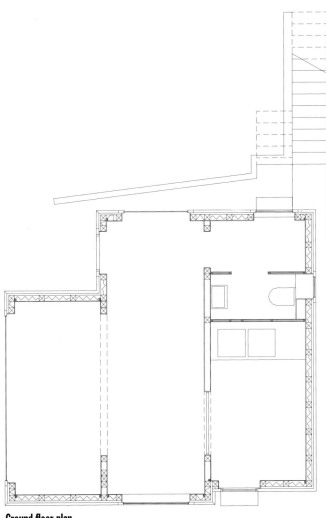

Ground floor plan

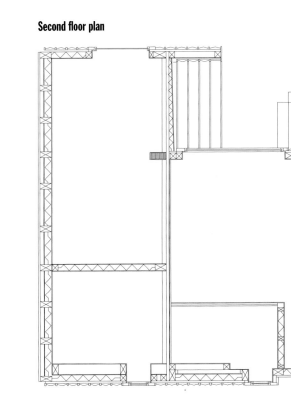

Cross-section B-B

First floor plan

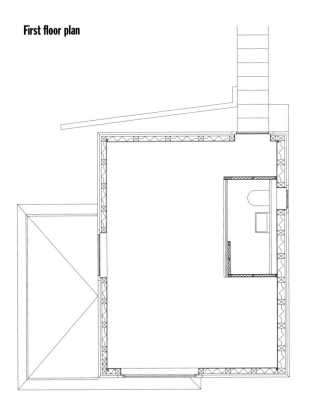

Second floor plan

The interior spaces are transparent and have excellent natural lighting provided by the windows. The birch wood modules conceal the kitchen area and the bathroom.

Construction detail of the window

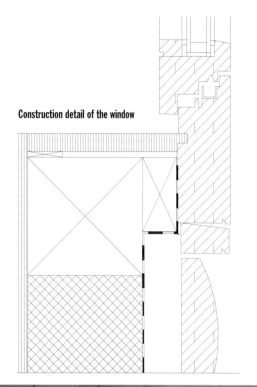

Earthship Biotecture
Earthship residences
Taos, New Mexico. USA

PHOTOGRAPHS: GALILEA NIN
DRAWINGS: SOLAR SURVIVAL PRESS

The Earthship has been designed to reduce our impact on the planet and increase our connection to it by utilising recycled and low embodied energy materials, passive solar heating and cooling, photovoltaic power system, catchwater, solar hot water, grey water and black water treatment systems. The techniques of building Earthships are presented and developed in a way that will allow Earthship owners to participate in and sometimes completely build their own home with many materials that are free byproducts of our existing society. The homes are designed with simplicity in mind. An owner builder with little or no construction experience could build or help build his or her own home. These units have been designed with many prefabricated components and these components can be delivered to the site and assembled by an owner builder. The owner's involvement in building educates them about the function of their home and empowers them to maintain and repair their home, making that home and lifestyle all the more sustainable.

The Earthship is a completely independent globally oriented dwelling unit made from materials that are indigenous to the entire planet. The major structural building components of the home are used automobile tires filled with compacted earth to form rammed earth bricks encased in steel belted rubber. These bricks and the resulting bearing walls they form are virtually indestructible. Used tires have low embodied energy; they are ready for reuse in construction without any alteration. Earth, another material with low embodied energy, is available on site or close to the site for ramming the tires and earth berming. Used cardboard is laid inside the tires to prevent the pounded earth from spilling out the bottom of the tire.

Non-loadbearing walls are constructed with aluminum cans, glass bottles and cement. Prefabricated building components can be mass-produced locally, saving time, resources and construction waste.

The three foot thick massive walls and the method of incorporating them into the earth create living spaces that retain a constant temperature. Thus, with solar gain and natural ventilation systems built in, this building will heat itself in the winter and cool itself in the summer without the use of centralised fossil or nuclear fuels.

These buildings produce their own electricity with a pre-packaged photovoltaic power system. This power unit is a pre-designed, pre-built "component" called the Power Organizing Module (POM). It can be installed and hooked up to by any electrician. Pre-designing reduces the need for specialists, therefore lowering the cost.

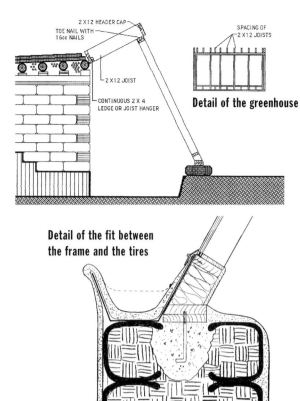

2 X 12 HEADER CAP

TOE NAIL WITH
16cc NAILS

2 X 12 JOIST

CONTINUOUS 2 X 4
LEDGE OR JOIST HANGER

SPACING OF
2 X 12 JOISTS

Detail of the greenhouse

**Detail of the fit between
the frame and the tires**

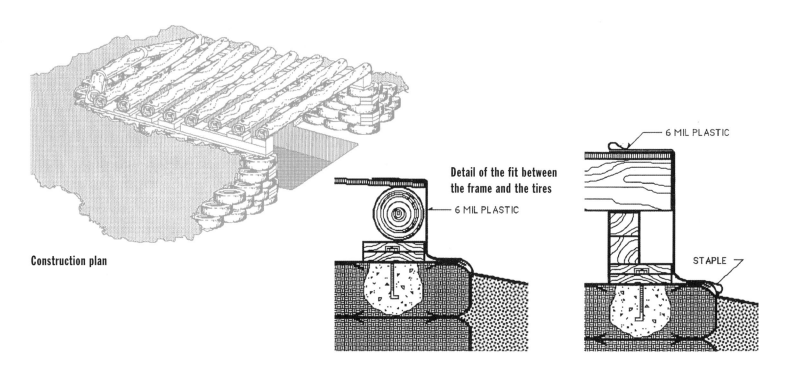

Construction plan

Detail of the fit between the frame and the tires

6 MIL PLASTIC

6 MIL PLASTIC

STAPLE

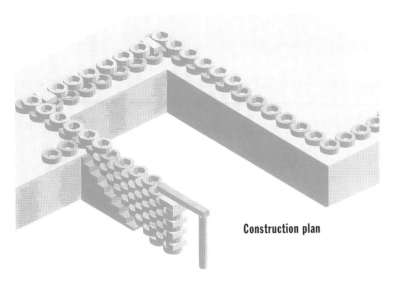

Construction plan

The use of atypical elements in the construction of these dwellings—such as glass bottles and aluminium cans—favours recycling. Due to their design and consistency, they may be used in different ways to create original and economic spaces.

Detail of the staircase

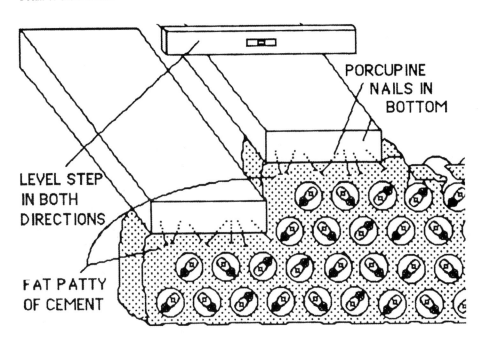

PORCUPINE NAILS IN BOTTOM

LEVEL STEP IN BOTH DIRECTIONS

FAT PATTY OF CEMENT

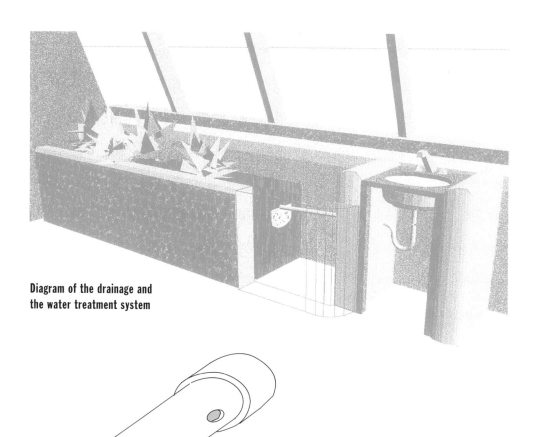

Diagram of the drainage and the water treatment system

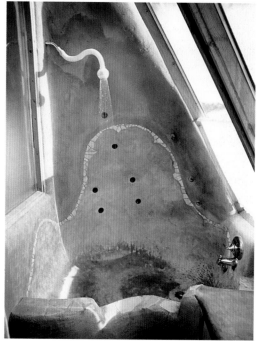

The buildings collect their own water from a unique metal roof, silt catch and cistern system and treat their own sewage through greenhouse technology that allows contained flush toilets.

Catch water systems elicit low water consumption and go hand-in-hand with grey water and black water treatment systems that cleanse and reuse water.

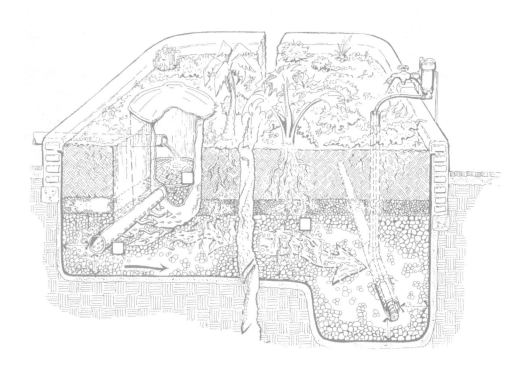

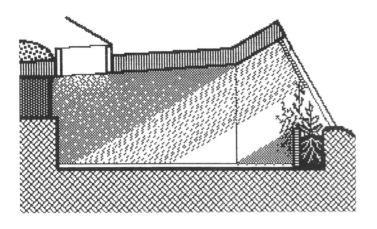

Lighting system

This rammed earth, passive solar structure has "designed down" mechanical systems, enabling its residents to live off of a relatively small solar power system. The idea is to incorporate the most efficient and low energy use equipment and appliances, reducing the overall electrical needs. There are no electrical heating demands because of the passive solar design. The glazed south sides of the building admit daylight, eliminating the need to use electrical lighting in the daytime.

Shigeru Ban
Paper House
Yamanakako, Japan

PHOTOGRAPHS: HIROYUKI HIRAI

Shigeru Ban, one of the most innovative architects of recent time, has been studying the characteristics of paper as a building element since the eighties Though there is some resistance to its use in a structural system, paper is a material that may be treated—like wood—to make it resistant to fire, water and damp. It is also easy to recycle and economic. This allowed the architect to work with this element in designs that required speed and low cost, such as the pavilions for refugees from Ruanda and the provisional buildings to house the victims of the Kobe earthquake.

In this scheme, a dwelling of 110 m², the structure is formed by two square horizontal planes with a side of 10 metres and paper tubes aligned in an S-shape with a height of 2.7 metres, a diameter of 280 mm and a thickness of 15 mm. These tubes support the house and defined its different functional spaces, relating them to the surrounding landscape. Ten of these tubes support the vertical loads and eighty interior tubes support the lateral loads. The circle formed by these eighty tubes defines the living room, whereas the circle formed by the square defines the bathroom of the dwelling.

The separation from the exterior is created using a glass wall that may open or close and that can also be covered by canvas curtains to provide privacy and good insulation. The spatial continuity between the interiors and the landscape is achieved through the horizontal elements and the use of very diaphanous joinery, and through the definition of the interior spaces with the minimum number of elements, following the example of the great architects of Modernism.

The paper tubes also allow the spaces defined to maintain a very subtle relation with the surrounding spaces, allowing in the light and views between them.

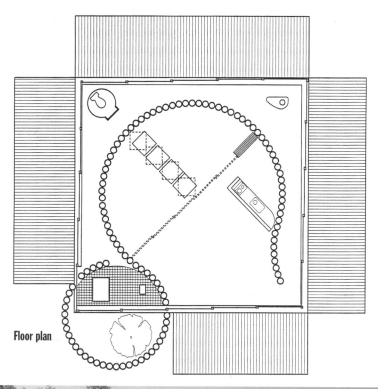

Floor plan

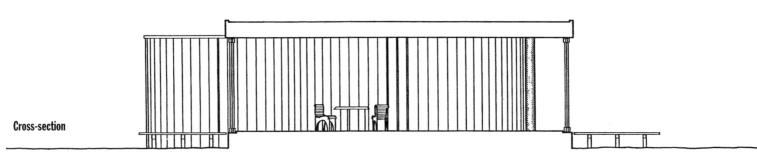

Cross-section

The relation between the interior and exterior spaces is a constant in the designs of Shigeru Ban. In this scheme, the enormous glazed window can be opened to leave the dwelling totally open to the exterior. To emphasise this relation even further, it was decided to create overhanging terraces that extend the floor area and take the dwelling closer to nature.

Bearth & Deplazes Architekten AG, Chur

Wohnhaus Willimann-Lötscher

Sevgein. Switzerland

PHOTOGRAPHS: RALPH FEINER

The four members of this family wanted a house that would be like the habitat of an animal, like the warren of a community of rabbits; in other words, they did not want a horizontal layout or an open flow between the different spaces. This premise served as the basis for a project for a house integrated into the landscape, with rooms laid out on half-floors. In a small clearing in the woods on the outskirts of the village of Sevgein, this house resembles a wedge inserted between the slope of the mountain and a spur of the hill, without altering the slightly inclined terrain on which it stands. This setting allows the house to open up to the panorama of the wooden valley of Vorderrhein.

In order to create generous spaces in spite of the small dimensions, there are two rooms on each floor, separated by a partition wall, the result being a vertical spiral of spaces. The entrance leads into a vestibule on two floors, which gives access to the dining room and kitchen on the lower floor and to the living room and studio on the upper floor. The four bedrooms are on the highest levels.

From the construction point of view, the house is constituted of a wooden structure of prefabricated elements, with skylights–also prefabricated–set in copper frames. In line with this range of standarized formats, two types of pivot window have been used, and these are found either separately or in combination in the various rooms.

The clients carried out a good deal of the work themselves (cladding of the façades with vertical boards, varnishing the wood, panelling of the interior walls with horizontal boards, painting), so it would be unrealistic to expect a high-precision professional finish in the detailing, corresponding to the degree of exactitude of the project drawings. Consequently, some of the design decisions relating to the overall appearance of the building proved decisive, such as those applied to the cladding of the façades: three types of board of different widths were chosen, to be positioned vertically, and these were also used to cover the window openings, thus producing a natural and harmonious balance between the panels on the façade, with contrast created by the texture of the boards.

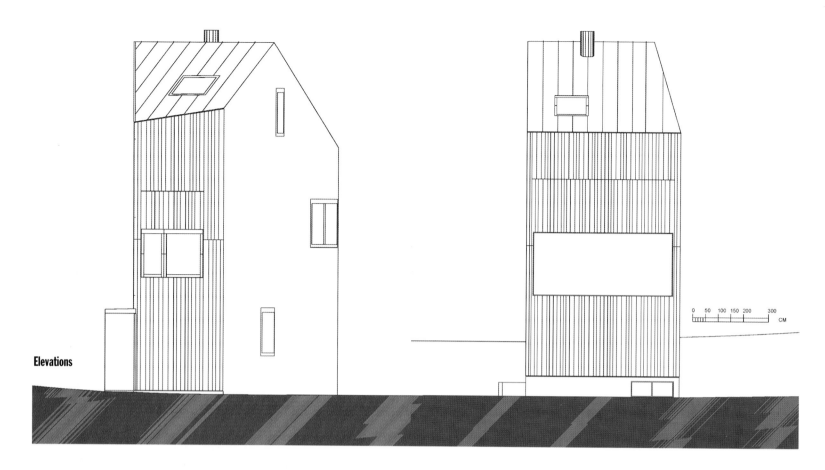

Elevations

Depending on the place from which it is observed, the dwelling appears semi-hidden behind the original design of its volumetrics.
The careful selection of wood used in the construction aims to demonstrate the integrating and rural nature of the new architecture.

Elevation

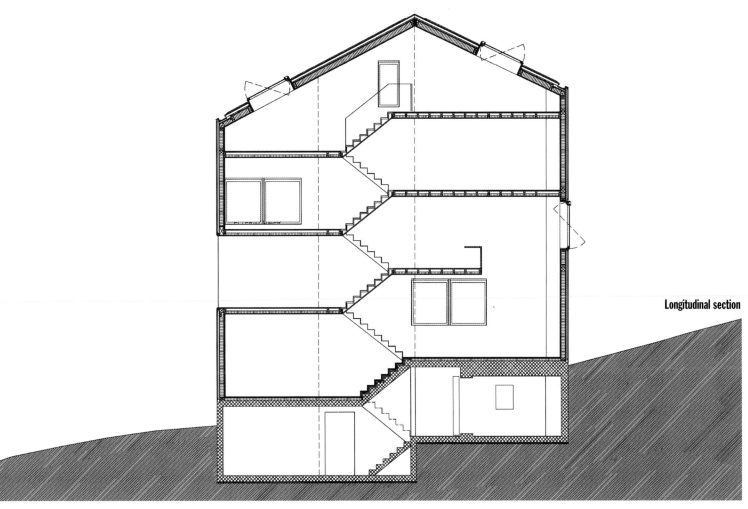

Longitudinal section

105

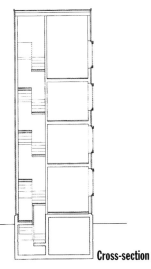

Cross-section

Conceived as a vertical structure supported by a gentle slope, this dwelling offers a succession of well-communicated spaces conceived in order to take full advantage of the views of the valley.

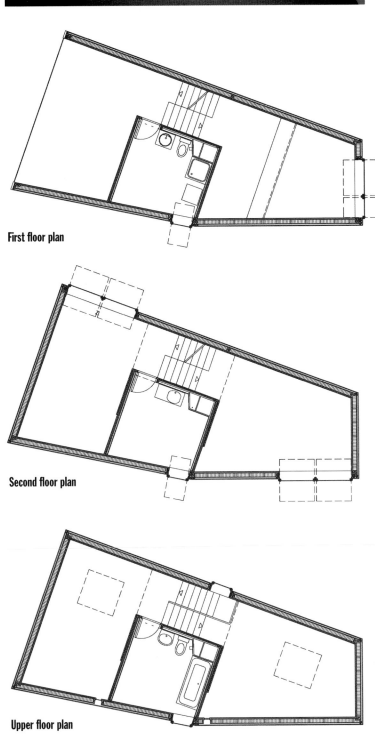

First floor plan

Second floor plan

Upper floor plan

Lake / Flato architects, Inc.

Bartlit Residence

Castle Pines, Colorado. USA

PHOTOGRAPHS: PAUL HESTER AND DAVID LAKE, FAIA

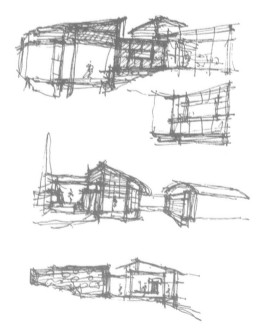

The Cliffs at Castle Pines is a site of exceptional natural beauty. Located on a prominent ridge at approximately 6,400 feet above sea level, the site has a commanding view of the entire "Front Range" from Pike's Peak to the south to Long's Peak far to the north, then drops 116' to the valley floor below. Mature stands of Ponderosa Pine, Scrub Oak, native grasses and flowers grow amongst the granite outcroppings and form the fragile natural ecology of the site.

The design solution took its cues from the natural land formations on the site. What appears to be a series of low granite riparian walls course down the hill and form the base of the walls of the house. Superimposed on this heavy base are the light exposed steel and glass "Pavilions" with their copper roofs that tilt to catch the sun and the views while sheltering the glass from the rain and snow. The house in partially dug into the western slope of the cliff and is barely visible from the entrance side of the house. A sod roof "meadow" covers the guest rooms and further merges the house with the natural landscape.

The materials for the house were chosen for their character, patina and enduring nature. Native Empire, Gunnison granite and Colorado gray-buff sandstone are used throughout the house both as the interior and exterior walls and paving. The interior walls that are not stone are a soft buff plaster or natural white oak. The custom wood cabinetry and floors are white oak. The exterior copper walls and roof will patina to a worn copper-penny color that merges with the rocks, tree trunks and soils of the site. While the exterior steel was finished against the elements, the interior steel was oiled to enhance the natural grain of the steel. The majority of the interior and exterior lighting is purposely kept to a minimum, with as many concealed or partially concealed sources as possible.

The "quarry" pool sits in a natural swale at the edge of the cliff. A grotto was carved out of the rock and serves as a sun-catcher and shelter from the wind.

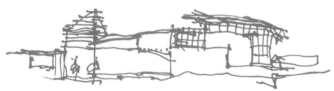

This dwelling was approached as an exercise of connection between the exterior and interior spaces so that its occupants could enjoy the excellent views offered by the site. This also provided easy access to the building.

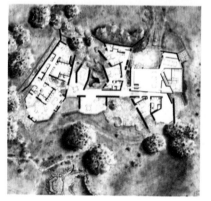

The stone walls appear to be a prolongation of the mountain within the dwelling. The use of local granite gives the scheme character and helps to regulate the indoor temperature naturally whilst respecting the environmental characteristics of the area.

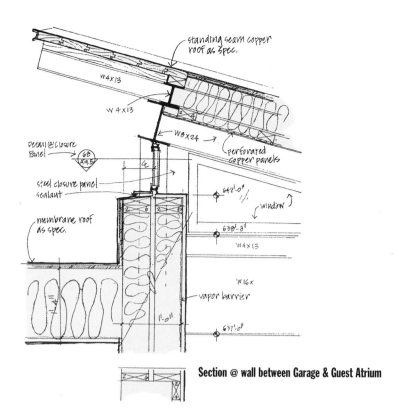

standing seam copper roof as spec.

W4X13

W 4 X13

Detail @ closure Panel

steel closure panel sealant

membrane roof as spec.

W8X24

perforated copper panels

64'-0"

window

638'-8"

W4X13

'N 16X

vapor barrier

11'-0"

637'-0"

Section @ wall between Garage & Guest Atrium

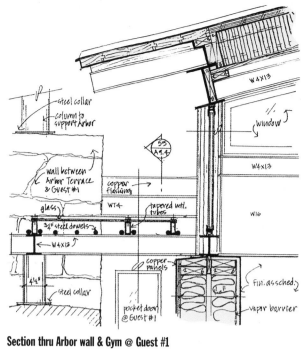

steel collar

column to support Arbor

window

W4X13

wall between Arbor Terrace & Guest #1

copper flashing

glass

WT4

tapered mtl. tubes

W16

3/4" steel dowels

W4X12

copper panels

4½"

steel collar

pocket door @ Guest #1

fin. as sched.

vapor barrier

Section thru Arbor wall & Gym @ Guest #1

The dialogue between the opacity of the stone and the brightness of the steel and glass is repeated in all the spaces of the dwelling.

The main room, which includes the office, the living room, the kitchen and the central atrium, was designed to function as a separate wing, providing a level of privacy and intimacy for the owners.

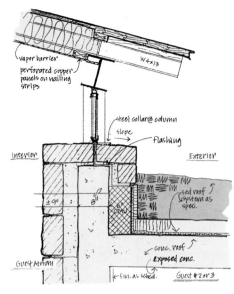

Section thru beam @ Guest Atrium/Gallery

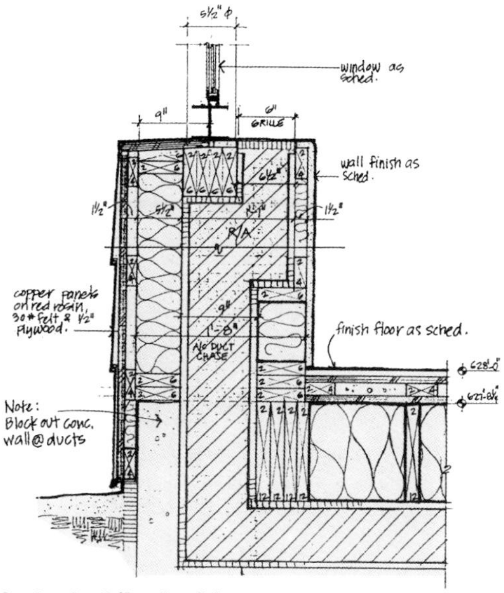
window as sched.
9"
6" GRILLE
wall finish as sched.
1½"
R/A
copper panels on red rosin, 30# felt & ½" plywood.
finish floor as sched.
NO DUCT CHASE
Note: Block out conc. wall @ ducts

Section thru R/A wall @ living room

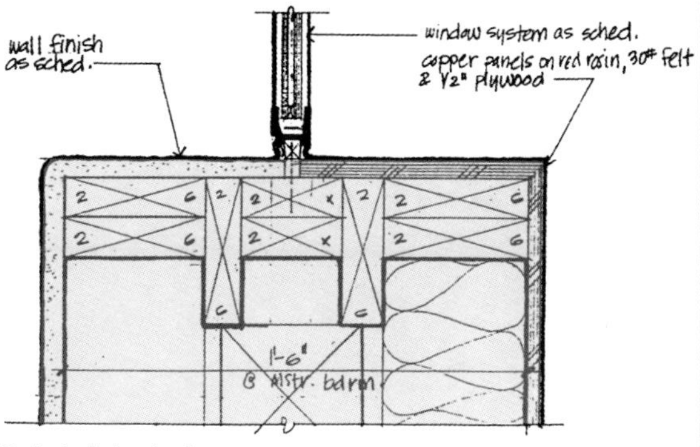
wall finish as sched.
window system as sched.
copper panels on red rosin, 30# felt & ½" plywood
1'-6"

Typical window jamb

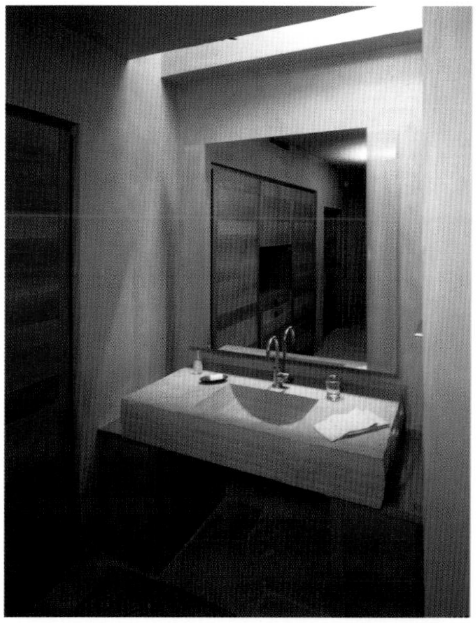

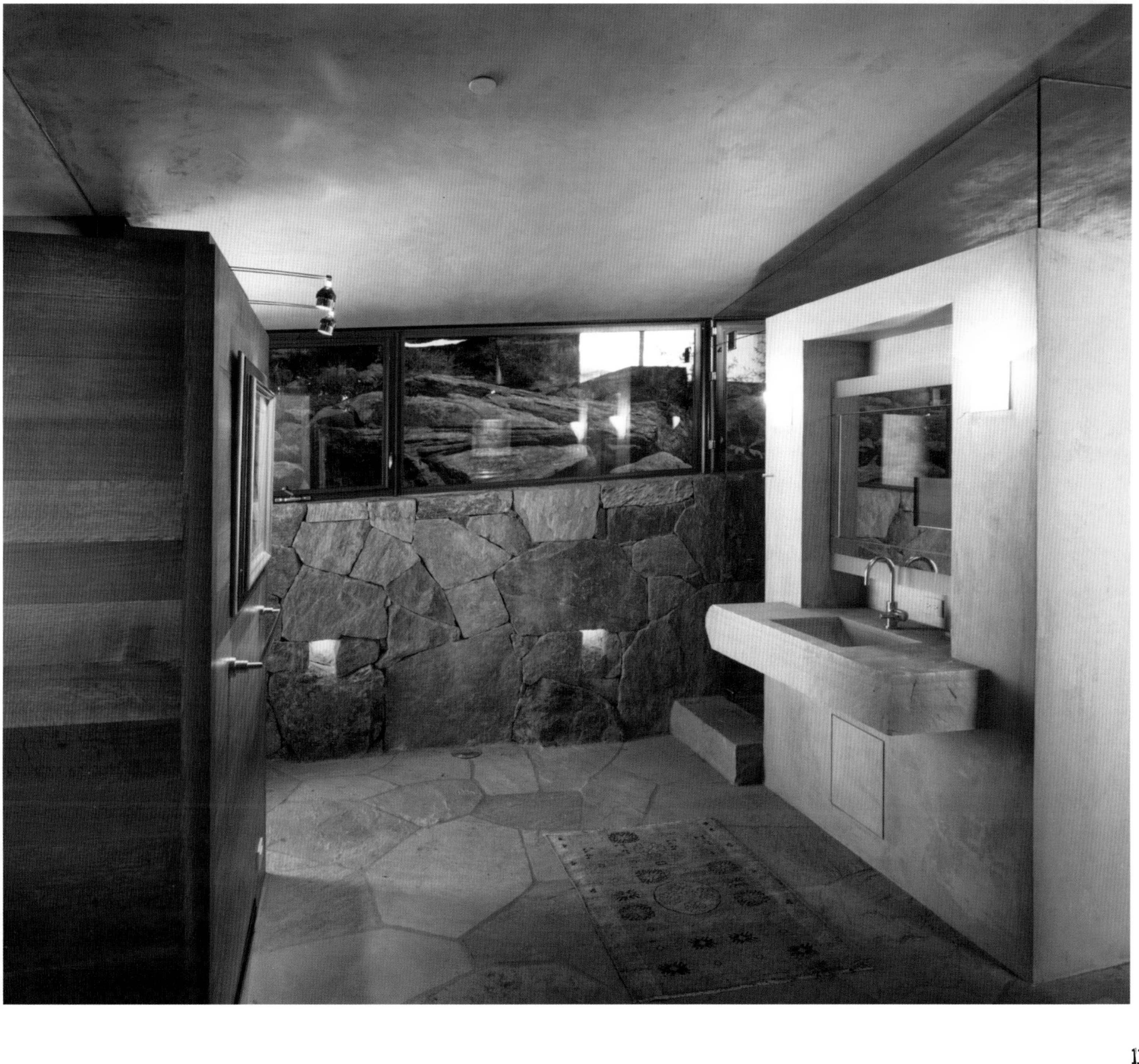

Eugene Tsui

The Watsu Center International School and Residence

Harbin Hot Springs, Middletown, California. USA

PHOTOGRAPHS: DR. EUGENE TSUI

On a majestic hillside overlooking the forested ravine and mountains of northern California sits a most unusual building. It is a special school of massage that is performed in water. Harold Dull, the inventor of this system, has built a residence and school which features five—8-meter to 10-meter diameter spheres formed in wood and sheeted in stucco cement and water-proof vinyl composite. Two outdoor pools are seemingly connected by a waterfall with a cave behind it. The waterfall helps to oxygenate recycled water through the house and school. The spheres contain classrooms, commercial kitchen, offices, sleeping quarters, residential quarters and a three-storey studio. Fireproofing is created by a series of outdoor sprinklers at the central roof of each sphere. In the event of fire the spheres are immediately immersed in water, becoming impenetrable to fire. At the base of each sphere is a water trough of moving cold water that naturally cools the air around the sphere. Small tubular openings contain solar-powered fans that draw the water-cooled air in and push hot air out the roof top openings, creating a natural, continuous air exchange system. Passive vacuum tube solar heating tubes heat water from the sun for all outdoors pools and indoor water use. Non-toxic, recycled paper pulp cellulose is sprayed on the ceilings of all spaces for sound absorption and sunlight reflection mitigation. It also has an R insulation factor of R-40.

Recycled paper pulp mixed with water-based glue is sprayed into the open structural cavities of the walls to provide excellent insulation. The spherical shape—minimizing surface area while masimising volume covered and structural strength—has proven itself to maintain a cool interior temperature even when the outside temperature reaches beyond 42 degrees Celsius and even without the insulation put in. The spherical shapes also activate wind flow to further cool the buildings. They are also extremely durable and stable, and easily able to resist earthquake forces acting upon them. In fact, it has been proven that the sphere is the strongest, most efficient shape in the universe. Every detail of the building is unique, including the stone retaining walls in which the rocks seems to grow straight out of the wall with brilliant glass marbles placed between the stones. The floors too are made of silver quartzite flagstone with flattened glass marbles of iridescent and transparent quality shimmering in the floors. This is a one-of-a-kind building open to public tours and visited by millions of persons from around the world.

The vein-like elements on the exterior of the spheres are hollow tubes which have cool air blown into them for summer cooling and hot air blown in for winter heating—much like a living organism—the structure adapts to the changing requirements of the environment and the utilitarian and spiritual needs of the people within.

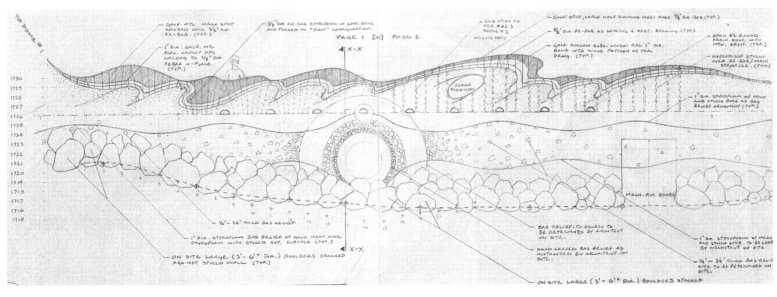

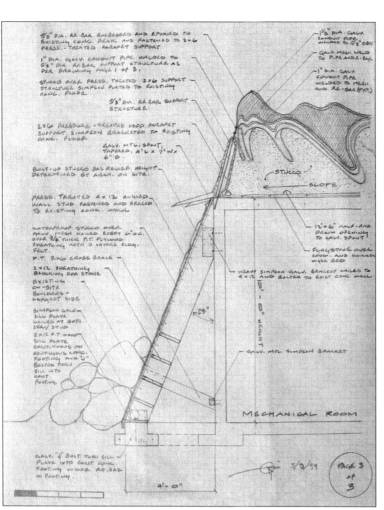

In the structural cavities of the walls, paper paste mixed with water-base glue was injected to provide excellent thermal and acoustic protection.

In addition to being light and non-toxic products, these materials can be used to create curved and organic forms.

Eugene Tsui

Ecological House of the Future

Bao An, Shenzhen, People's Republic of China

This 200 square meter apartment is a glimpse of the future of ecological apartment environments in China. The general ambiance is one of living in a natural setting surrounded by the curving and undulating forms of natural wind-swept stone interspersed with moving waterfalls, waterponds and abundant plants. But there is much more than meets the eye. The environment sustains itself with minimum maintenance. It is a living, interconnected system of self-regulating sewage treatment, clean water and air recycling, natural breeze control and self-activating windows that automatically adjust the amount of entering sunlight.

Exterior photovoltaic solar panels convert sunlight into electricity, which is stored for home used. Roof windmills also convert wind to direct power for electrical home use.

In sum, the future ecological house of China will be a place as beautiful and spiritually powerful as the forests, mountains, lakes, deserts and oceans. It is a place where nature and humanity come together for the benefit of both; a rejuvenating place where a person can come closer to the technological wisdom and poetic power of nature. A home of robust physical health, comfort and creative inspiration.

PHOTOGRAPHS: DR. EUGENE TSUI

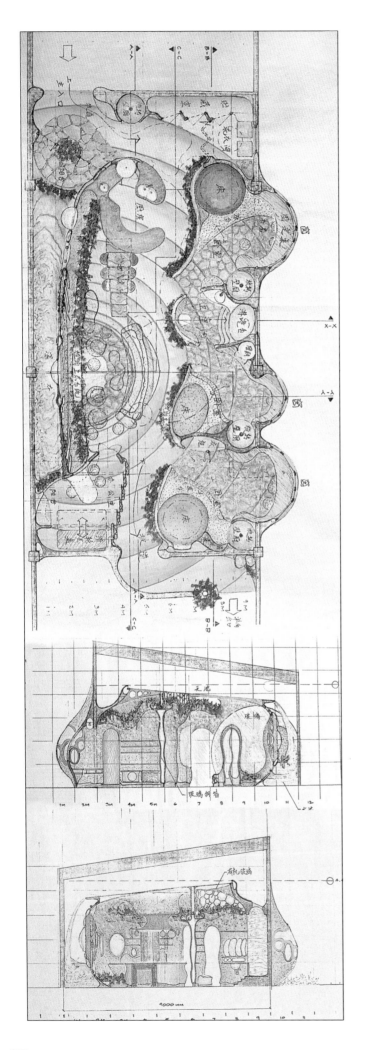

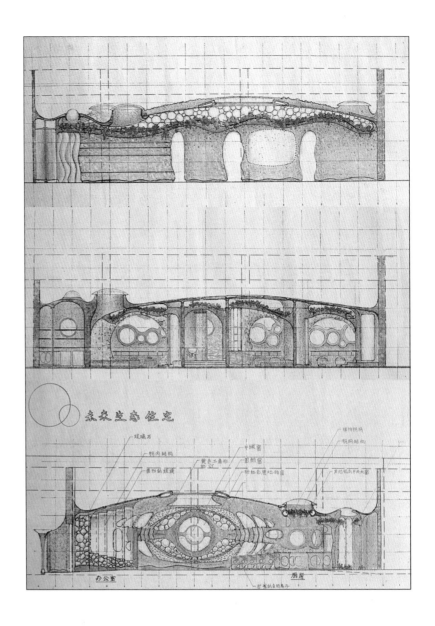

All exterior windows are clear hemispheres that allow viewers to see the left and right of the building's exterior and provide geater sunlight and wind resistance economically.

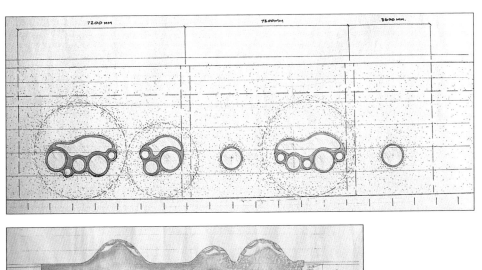

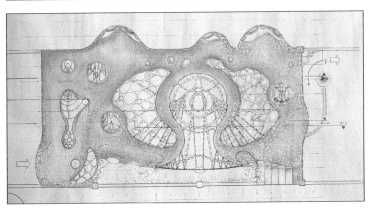

William McDonough + Partners
Virginia Beach Residence
Virginia Beach, Virginia. USA

PHOTOGRAPHS: PRAKASH PATEL

Situated on the shore of an environmentally sensitive cove, this residence for a large family is equipped with facilities for frequent entertainment. Constructed primarily of sustainably harvested wood, the house enjoys extensive views of Lynnhaven Bay, a constructed wetland pond, and preserved specimen vegetation.

A central pavilion joins the house's north and south wings, clearly defining its public and private spaces. This separation, the landscaping, and the poolhouse create clear zones of occupation, using the residence's walls to capture garden spaces and to blur the boundary between indoors and out. The central pavilion, which houses the kitchen and family room, serves as the transition between these interior and exterior landscapes, tracing the paths between them. An occupiable grass roof above the pavilion doubles as an terraced patio, elevating and extending the garden theme.

Thanks in large measure to the enthusiastic support of a local supplier, the vast majority of lumber is sustainably harvested or certified by the Forest Stewardship Council. The extensive use of wood both inside and out connects the house's modern lines to its native setting, which includes landscape plantings of native species of trees, shrubs, and grasses that complement the preserved specimen pines, black locusts, and white, red, and live oaks.

The exterior makes use of certified Spanish cedar ship-lap siding, plywood veneer, and benches, which are set off by a certified "Billy Webb" trellis and wood deck. Inside, an asymmetrical curved ceiling of tongue-and-groove cherry admits abundant amounts of light through the roof's clerestory windows, which will deepen the rich tones of the photosensitive cherry floors and millwork over time. Clear finishes preserve the wood's warmth and texture, reinforcing the natural character of the domestic setting.

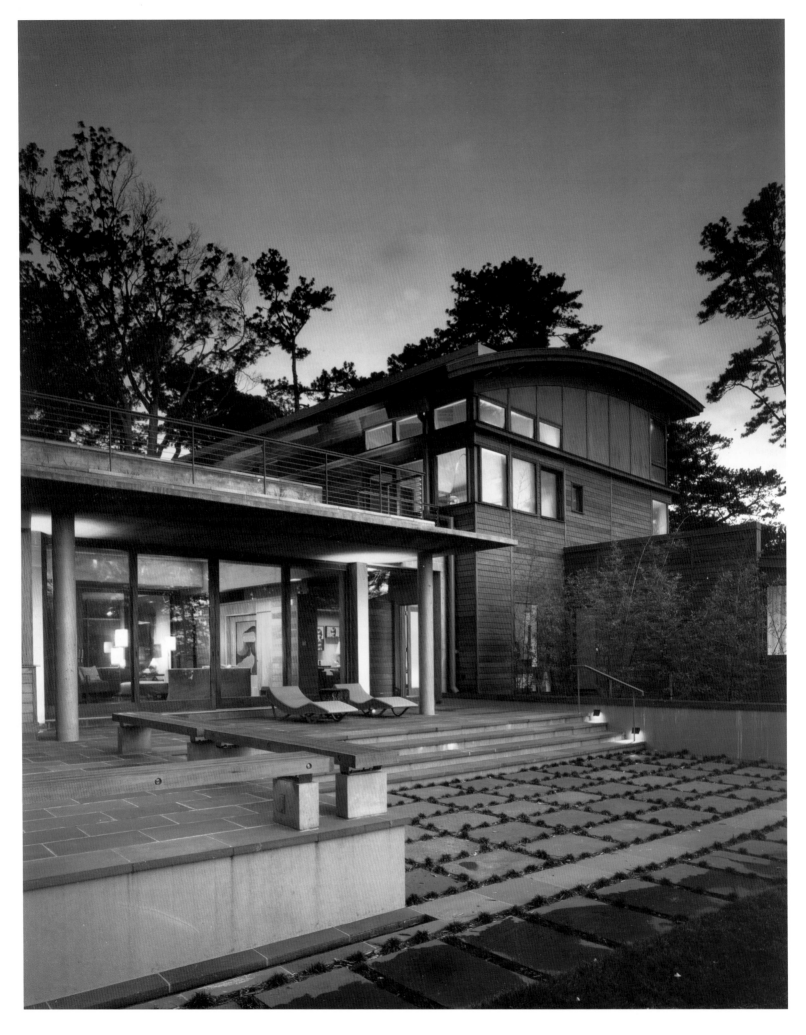

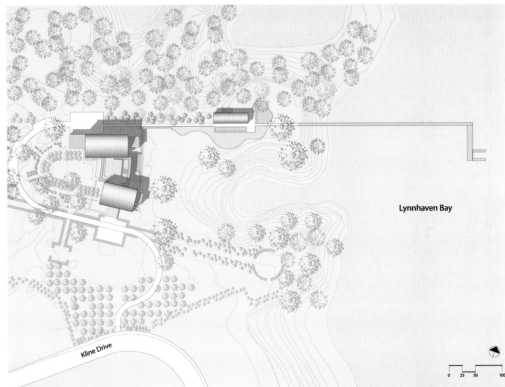

Lynnhaven Bay

Kline Drive

0   25   50   100

For the construction of this dwelling only wood from local trees that were easy to reproduce was used. Thus, the scheme shows its respect for the surrounding nature.

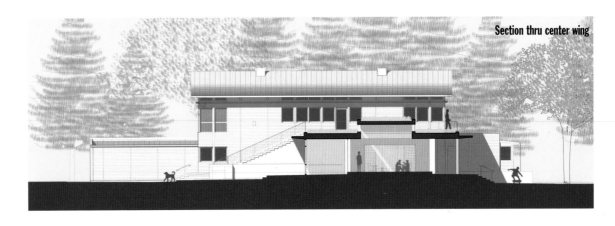

The arrangement of the different buildings of the dwelling highlights the presence of the enormous garden. In addition to offering a gratifying visual experience to the occupants, this garden was designed bearing in mind the biodiversity of the area so that the scheme helps the conservation of protected species instead of hindering it.

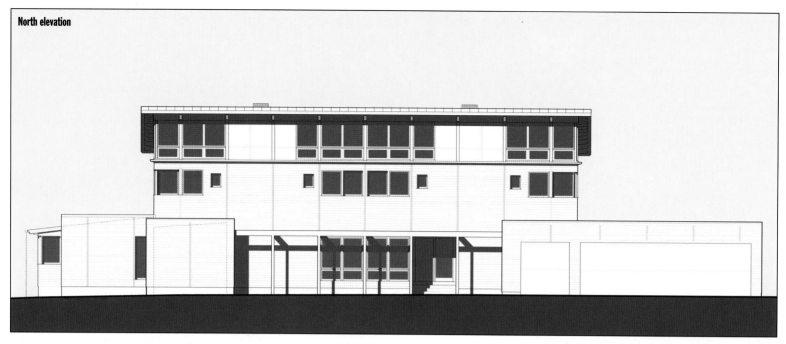

North elevation

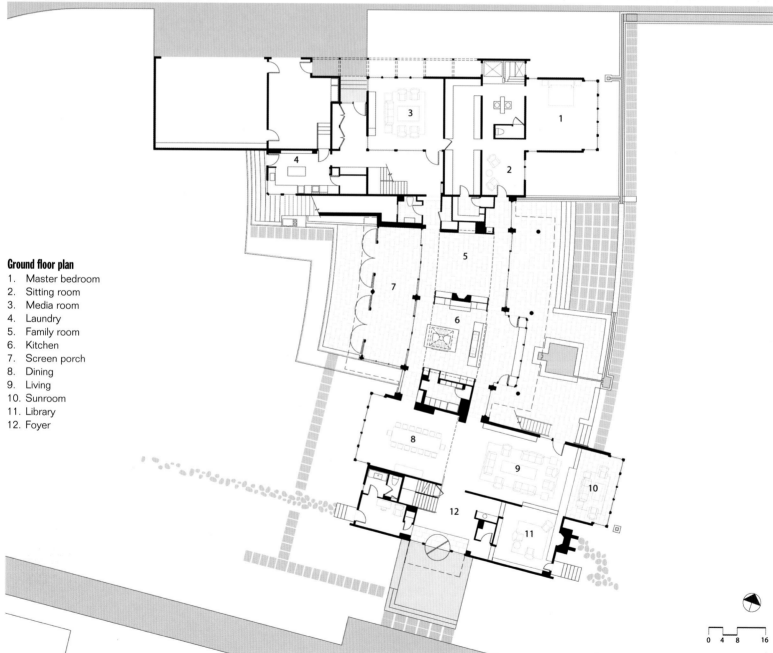

Ground floor plan

1. Master bedroom
2. Sitting room
3. Media room
4. Laundry
5. Family room
6. Kitchen
7. Screen porch
8. Dining
9. Living
10. Sunroom
11. Library
12. Foyer

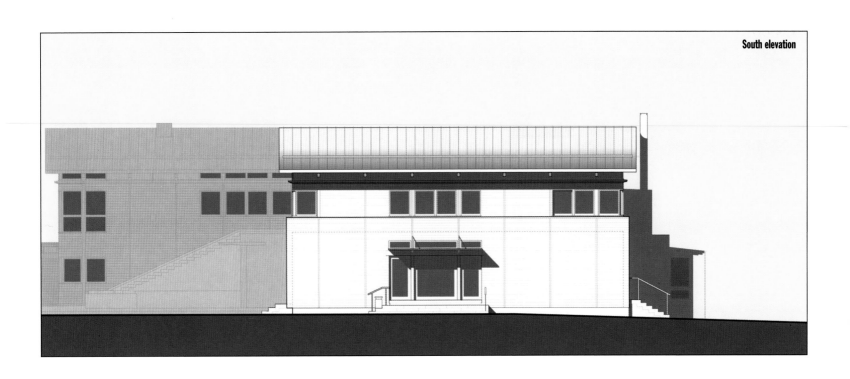

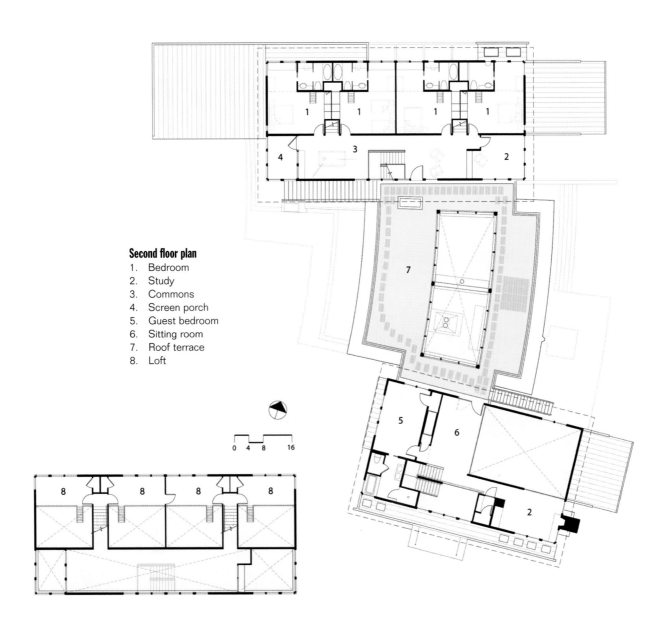

Second floor plan
1. Bedroom
2. Study
3. Commons
4. Screen porch
5. Guest bedroom
6. Sitting room
7. Roof terrace
8. Loft

0 4 8   16

Simón Vélez
Casa Mercedes
Manzinales. Colombia

FOTOGRAPHS: SIMÓN VÉLEZ

This structure is part of a larger set of buildings that are all part of the same house, a hacienda for a horse ranch.

This was the first all-bamboo structure designed by Velez about 15 years ago. The joinery system he developed here has allowed much larger structures to be built since then. The secret is to use mortar to fill the hollow of the bamboo where the bolt passes through to join two pieces of bamboo. The bamboo used in this structure comes from the same ranch and was untreated. Traditionally, in this region, bamboo was not treated and there are buildings over 100 years old that use traditional joinery, usually ropes.

Velez collaborated on the two new ideas that allow bamboo structures to be literally stronger than any other material: the joinery and the recently developed system of treatment that uses smoke to discourage insects and rot. In Japan, the bamboo rafters of farmhouses have lasted for hundreds years with no damage because of the cooking fires on the floor below.

When the Spaniards arrived to this region, as far as the eye could see was the native guadua bamboo, 20 metres tall and 0,5 metres apart. The farm-

ers soon learned that the most fertile ground they had ever seen was under the guadua. 300 years later, guadua is still the cheapest and easiest material which to build. The earthquake in the Quindio region killed nearly 100 people and destroyed 75% of the buildings. All the people were killed by falling concrete and brick. None of the new bamboo structures collapsed.

About 100 years ago, some major fires in large Colombian cities demonstrated that bamboo has about the same resistance to fire as wood. Because there were no fire trucks, fire hydrants or firemen to fight the fires, they caused huge damage and a prejudice against building with bamboo was developed. More recent practices such as plastering over the bamboo in the walls and using tile for roofs ensure that fire does not spread from one building to another, and since the earthquake, there has been a very inspiring resurgence in bamboo construction in the region and in neighbouring Ecuador.

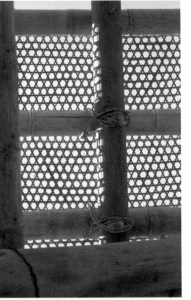

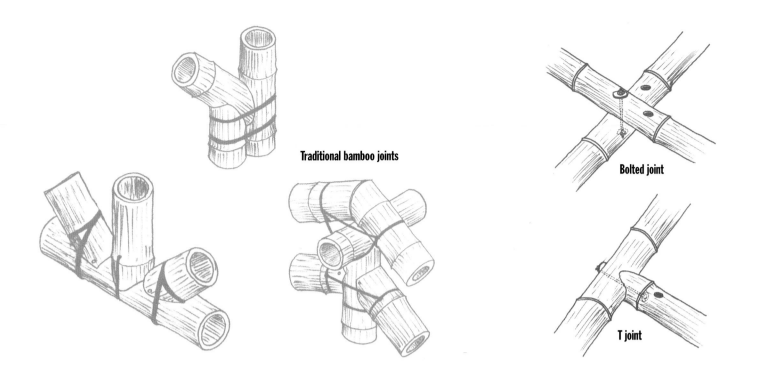

Traditional bamboo joints

Bolted joint

T joint

The drawings on the left show traditional ideas of joints that will collapse when one tries to make larger structures. Any "lashed" joint will work loose due to the changes in the humidity and changes in the diameter of bamboo. Usually, is it too difficult to constantly make the joints strong again. That is why all traditional bamboo structures are limited to very small spans.

The drawings on the right are an example of the system of joinery used in all of Simon Velez's buildings. Bolts and mortar are used to fill the hollow of the bamboo where the bolt passes through to join two pieces, and with this system it is possible to build bigger and stronger structures.

Polly Cooper & Ken Haggard
Trout Farm Complex
San Luis Obispo, California. USA

**PHOTOGRAPHS:POLLY COOPER
KEN HAGGARD
JOSEF KASPAROWITZ**

One of the keys of this scale of planning and design is the concept of the watershed. This site contains part of four interested watersheds.

The largest is the Salinas River Watershed, which is almost totally contained on the project site.

The programme is a mixed use architectural complex of offices, research and facilities that draw its energy for heating, cooling, lighting, and electrical generation from the site by passive means. 80% of the wood for construction was obtained from the site as well. In addition, there are the requirements of healing the site and watershed due to previous mining, grading, and motorcycle damage done from 1910 to 1970. Sporadic wildfires are also a major design factor for this location.

The Cooper & Haggard power system is a hybrid solar / hydro DC to AC power system. The hybrid nature of this system allows this family to use seasonal advantage to produce power from solar energy in high sun periods, and from hydro-power during the rainy season.

Viable water flow exists year round here, but is of course best during the rainy season. Use of the water flow is optimised by the exchanging of nozzles within the turbine.

SOLAR DESIGN FEATURES: 1. South facing low-e windows with optimised overhangs and wing walls. 2. Double glazed skylight with skylid movable insulation. 3. Concentrated mass of concrete block used as shear walls and 9-inch (23-cm) thick water tanks with selective surface paint below south facing windows where exposed to direct winter sunlight. 4. Distributed mass of floor slabs with perimeter insulation and 1-1/2-2 inch (3, 8-5 cm) interior stucco on straw bale walls. 5. Designed so that only natural light is used during daylight hours. 6. 4 kW solar electric system.

7. Solar water heating system with high efficiency propane backup.

ENERGY EFFICIENCY FEATURES: 1. Highly insulated ceilings of various types (straw bales, aircrete and blown–in dry cellulose). 2. North, east and west walls rice straw bales on edge. 3. Insulating shades. 4. Movable shade cloth sunshades on some east and west locations. 5. Shading trellis of recycled pipe from site

ENVIROMENTAL / HEALTH FEATURES: 1. Use of dead trees on site killed by the fire for most timber (Sargent Cypress and Douglas Fir – structure; Coulter Pine – ceilings; White Alder – cabinets and trim; Sycamore – special cabinets; Sargent Cypress – selected furniture; all the above and California Bay – trim). 2. Compact fluorescent for all electric lighting. 3. Sun forest refrigerator, Staber 2000 clothes washer. 4. Energy-efficient printer.

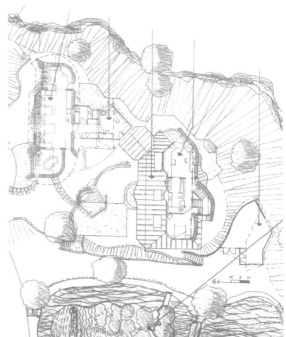

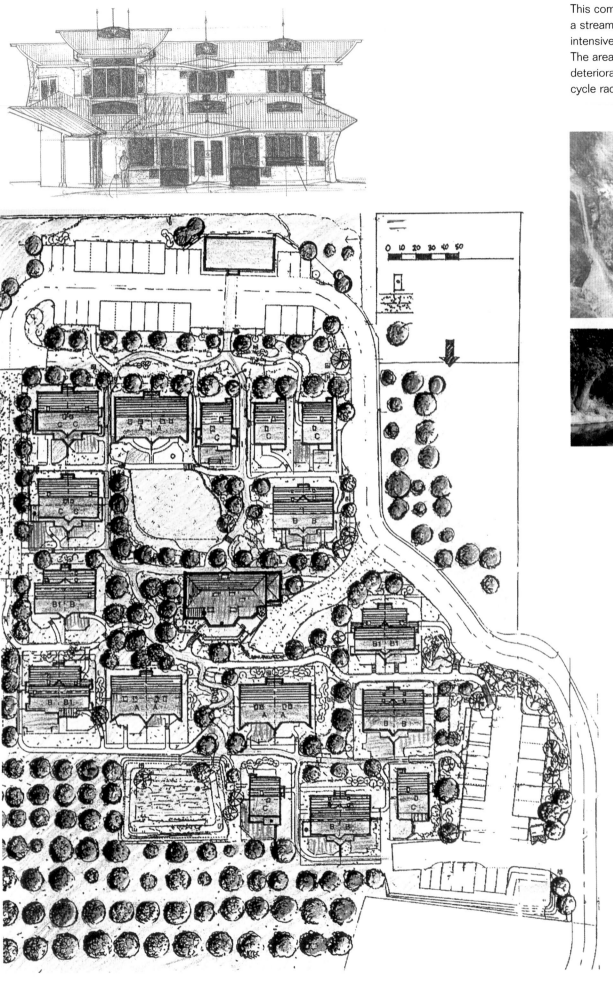

This complex is located in the basin of a stream that runs through a small, intensively-farmed Californian valley. The area was improved after years of deterioration due to mining and motorcycle racing.

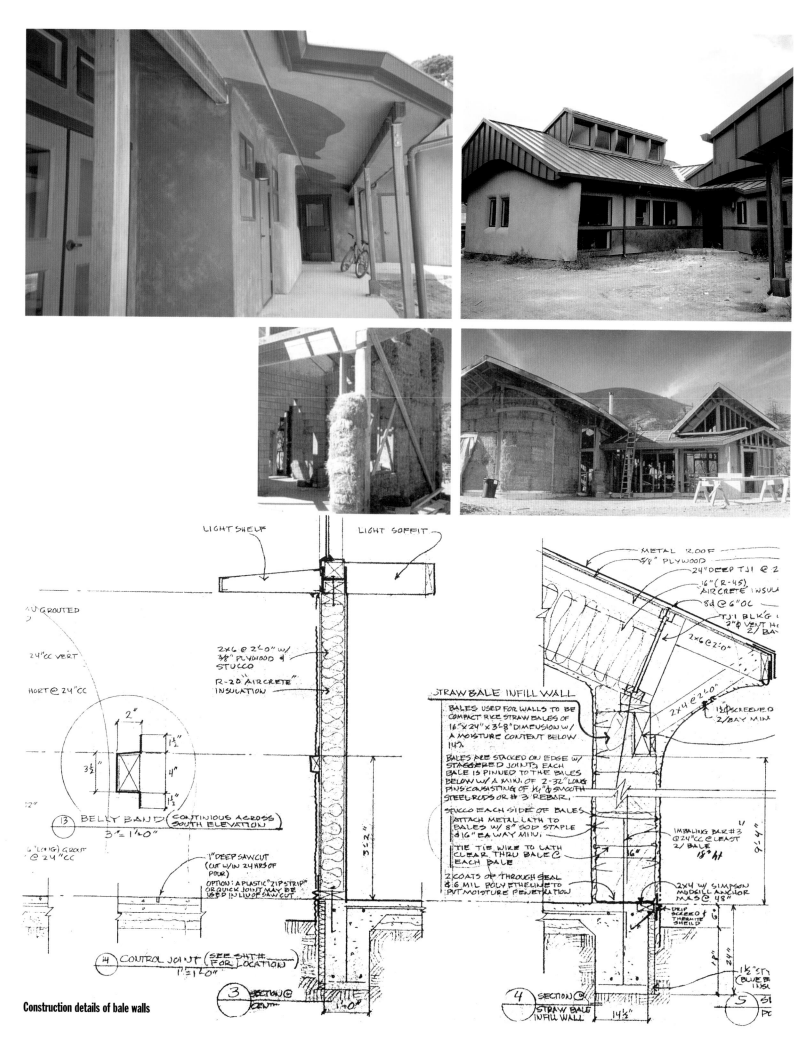

Construction details of bale walls

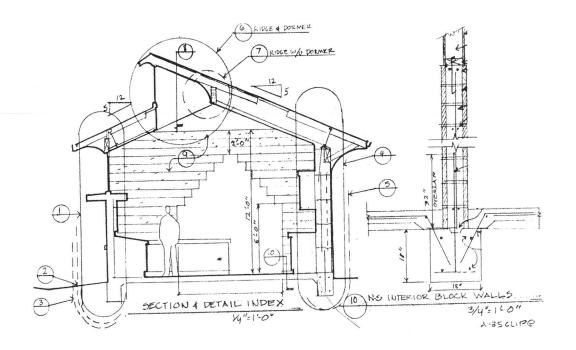

SECTION & DETAIL INDEX
1/4"=1'-0"

6 RIDGE & DORMER
7 RIDGE W/O DORMER
10 N·S INTERIOR BLOCK WALLS
3/4"=1'-0"

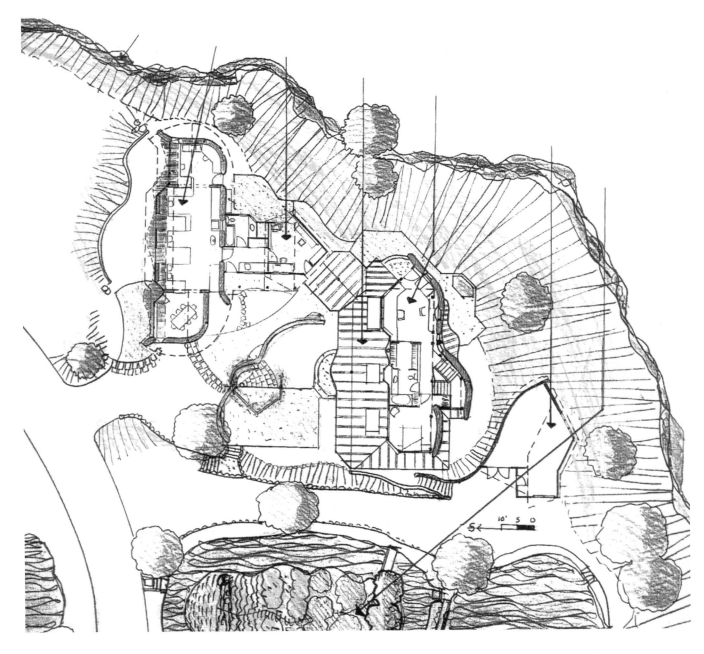

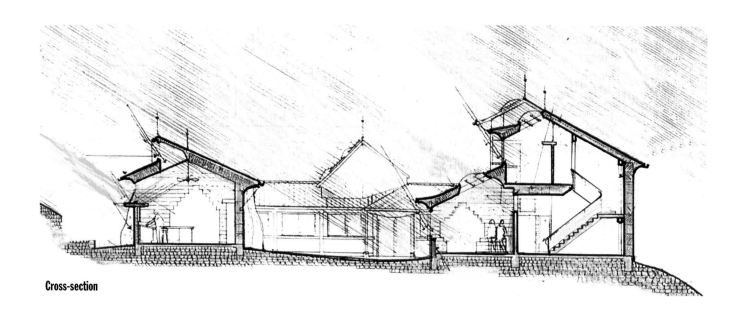

Cross-section

Apart from the thermal protection provided by the bales of straw, the scheme aimed to maximise the influence of sunlight on the building. The solar panels and the openings take advantage of this rich source of natural energy.

Jones Studio, Inc
Walner Residence
Scottsdale, Arizona. USA

PHOTOGRAPHS: TIMOTHY HURSLEY

The shape of the home—a parallelogram built on two levels—results from key elements offered by the site. These include a boulder pile in the foreground, the western exposure, views of distant city lights, and the relationship of the site to the street. Behind the house the 85-foot long negative edge lap pool echoes the shape of the house.

Like a giant bird poised for flight, the home appears ready to soar above the desert in far north Scottsdale, Arizona. The home is clad in quartz coloured titanium zinc roof and wall panels. The stretched out roofline combined with the cantilevered steel structure of the main level effectively floats the house above the desert. The drama of the super-extended roof edge began with the practical need to shade the high percentage glass walls.

At the upper level, the west windows incorporate mitered-glass corners, like a fish aquarium, providing uninterrupted views. Multiple areas for relaxing and dining allow for flexibility and accommodate the owner's extended family. Custom clear maple cabinetry accents the kitchen and houses the media center. The aluminum grating of the deck reflects the evening illumination and is comfortably cool even on hot days.

The architect-designed carpet complements the space and is integral with the architect-designed circular cocktail table whose two legs pierce its glass top and culminate with candleholders. A suspended fireplace with its pleated-glass curtain can be viewed from inside the house or out.

On the lower level a glass wall pivots open to extend the inside to the outside, providing ventilation and access to the lower deck and pool area, which is shaded by the cantilevered balcony above.

The 3800 square foot structure is constructed primarily of insulated and sandblasted concrete block, structural steel, glass, and titanium zinc. Titanium zinc is lightweight, easy to install, requires zero maintenance, and will last a lifetime. Also, because the house was designed to "float" above the desert, the architects did not want to use any materials (such as masonry) that would increase the weight of the cantilevered structure. Therefore, they utilized a light-weight skin that could patina naturally and blend with the desert surroundings.

A pan formed standing seam system accentuates form and profiles the folded wall / roof planes without the interruption of ridge flashing. The standing seams are continuous from roof to wall with individual knuckle caps at each joint. The flexibility and highly durable characteristics of solid zinc provide virtually unlimited opportunities for the creative and expressive architect and value conscious owner.

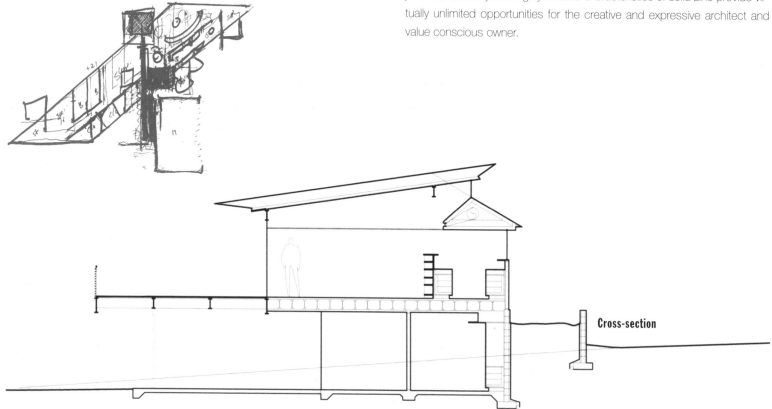

Cross-section

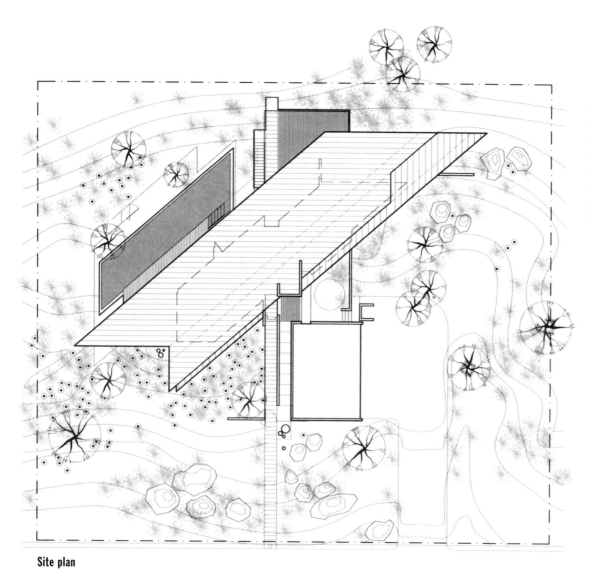

Site plan

The configuration of the dwelling —a parallelogram built on two levels— emerged in response to the natural elements of the location. Thus, the building stands between boulders and small trees, taking advantage of views of the large esplanades in the west and keeping a successful relation with the street.

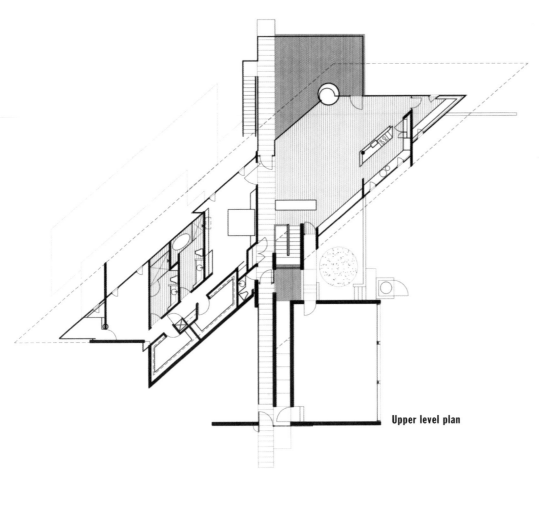

Upper level plan

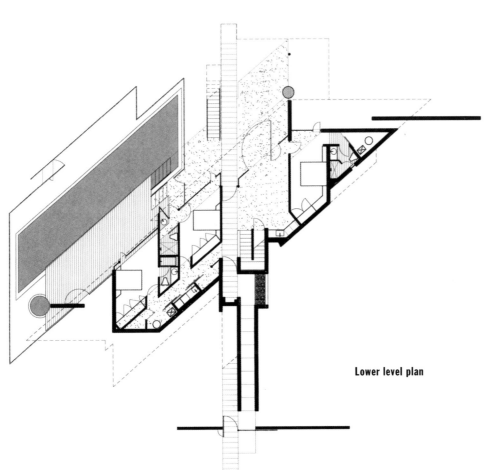

Lower level plan

Robert Oshatz

Gibson Boathouse/Studio

Lake Oswego, Oregon. USA

The Gibson's had an existing boathouse but felt it was a blemish on their property. They wanted to reuse the existing boat stall but build a new boathouse while adding a new studio and study. The site went from the lake up the hillside to the driveway above. Since the driveway to their property is shared with neighbours, it was decided to build the studio into the hillside and have a sod roof so the structure would not be noticed as the neighbour drove down—the driveway. Mrs.Gibson, an artist, wanted her studio space to have high ceilings and ample natural light. Mr. Gibson, an entrepreneur, wanted a more intimate space to keep track of his business activities. The structure grows out of stonewalls that are shaded by an arching sod roof. The roof is constructed with straight Douglas fir glue- laminated beams and fir decking.

PHOTOGRAPHS: ROBERT OSHATZ

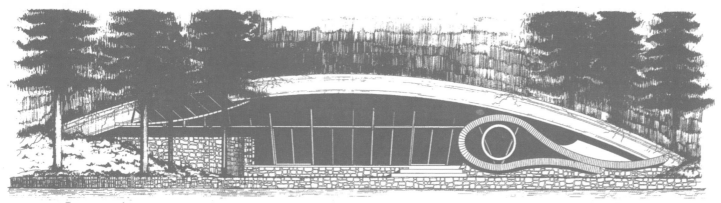

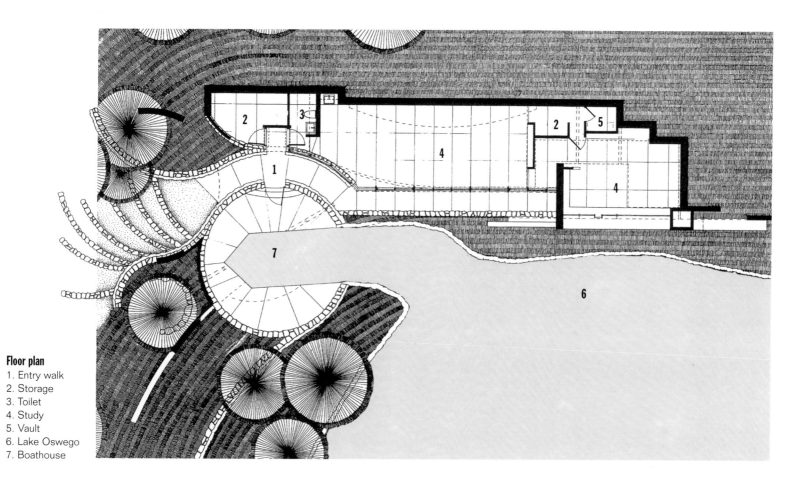

Floor plan
1. Entry walk
2. Storage
3. Toilet
4. Study
5. Vault
6. Lake Oswego
7. Boathouse

149

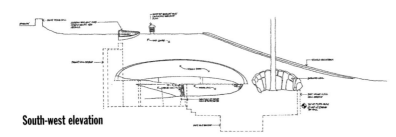

South-west elevation

The idea of building a studio in the jetty area added a privileged space to the dwelling. It is camouflaged by the vegetation-covered roof, which hides it from the passers-by who use the path leading to the other houses.

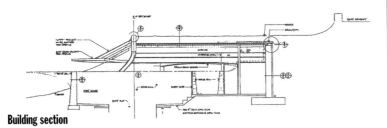

Building section

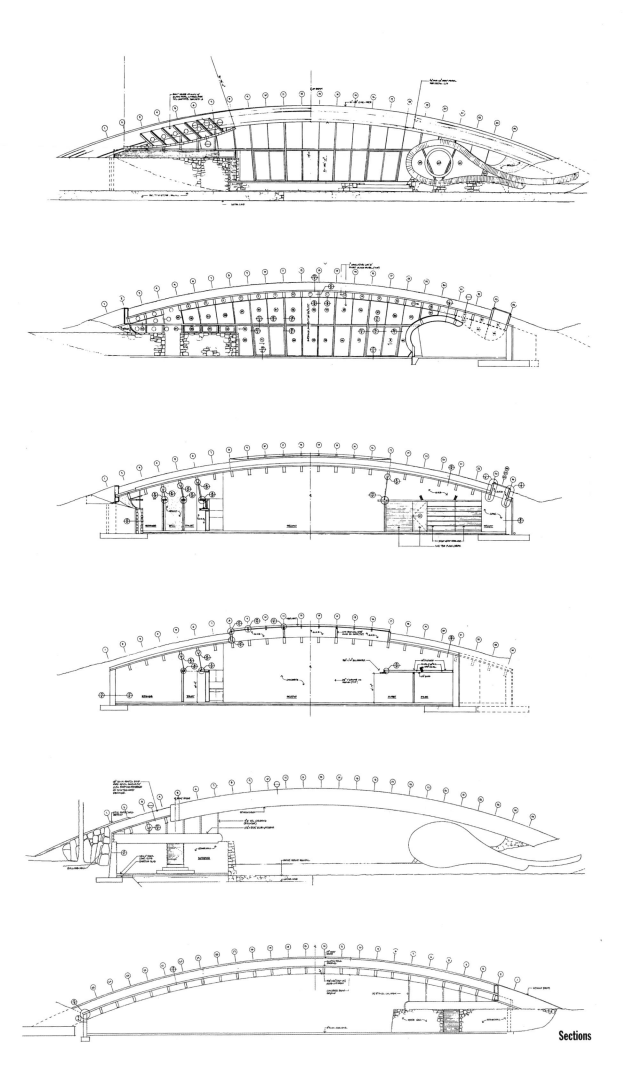

Sections

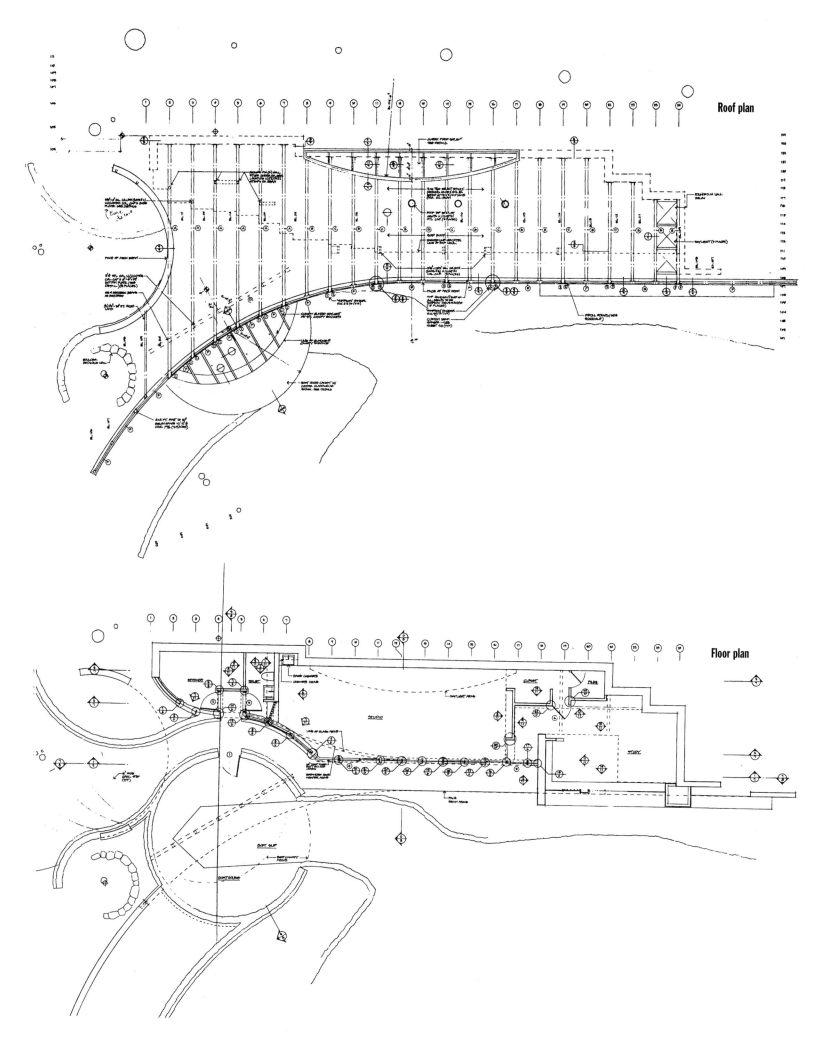

Roof plan

Floor plan

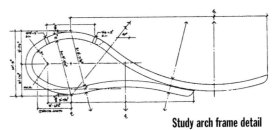

Study arch frame detail

The main elements used in this scheme—stone, wood and grass—are completely ecological and maintain a pleasant relationship with the surroundings of the studio-jetty. A small skylight in the vegetation-covered roof provides toplighting for the interior of the studio.

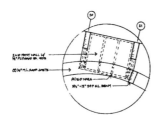

Pony wall from arch to roof

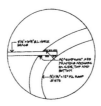

Transom to arch detail

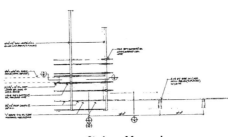

Study roof frame plan

Cottrell & Vermeulen + Buro Happold Consulting

Westborough Primary School

Westcliff-on-sea, Essex (England)

PHOTOGRAPHS: PETER GRANT PHOTOGRAPHY

The new school of Westborough offers a good lesson on the possibilities of using recycled material in construction. It is built mainly with recycled paper, though other materials such as wood are also used.

The school was designed by the architects Cottrell and Vermeulen together with Buro Happold, an industrial engineer who collaborated with SHIGERU BAN in the construction of the paper pavilion for the Expo of Hannover. This scheme was conceived to meet the challenge of demonstrating that paper is a useful material for the construction of permanent buildings. To achieve this, the building had to meet many strict measures: insulation, structural strength and fire protection.

To solve the structural problems caused by the tendency of paper and cardboard to creep under a continuous load, 11 tubes of paper were placed as pillars at each end. For a temporary building only four would have been necessary to support the weight of the roof. Though these tubes are one of the most obvious signs that the building is made of paper, the design of the roof and the drawings with origami by the artist Simmon Patterson make clear references to the qualities of the material. The roof suggests folded paper, and though the joints and drains were designed carefully, it was decided to use transparent plastic for weather protection and to allow daylight through the skylight. Fire protection was achieved by treating the paper with asbestos cement and using a type of recycled gypsum.

The cladding panels and the tubes used as pillars were made from recycled paper collected by the students of the school.

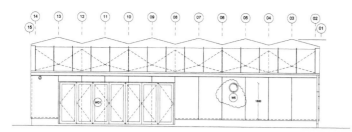

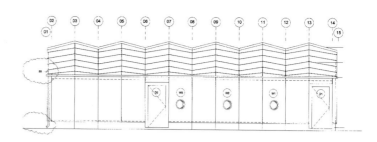

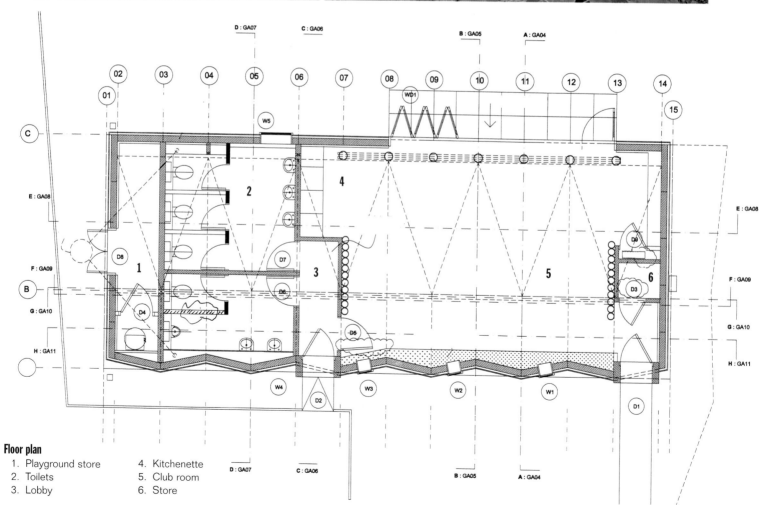

Floor plan

1. Playground store
2. Toilets
3. Lobby
4. Kitchenette
5. Club room
6. Store

The paper tubes that support the structure are perfectly aligned at each end of the building. They were treated with varnish to prevent staining.

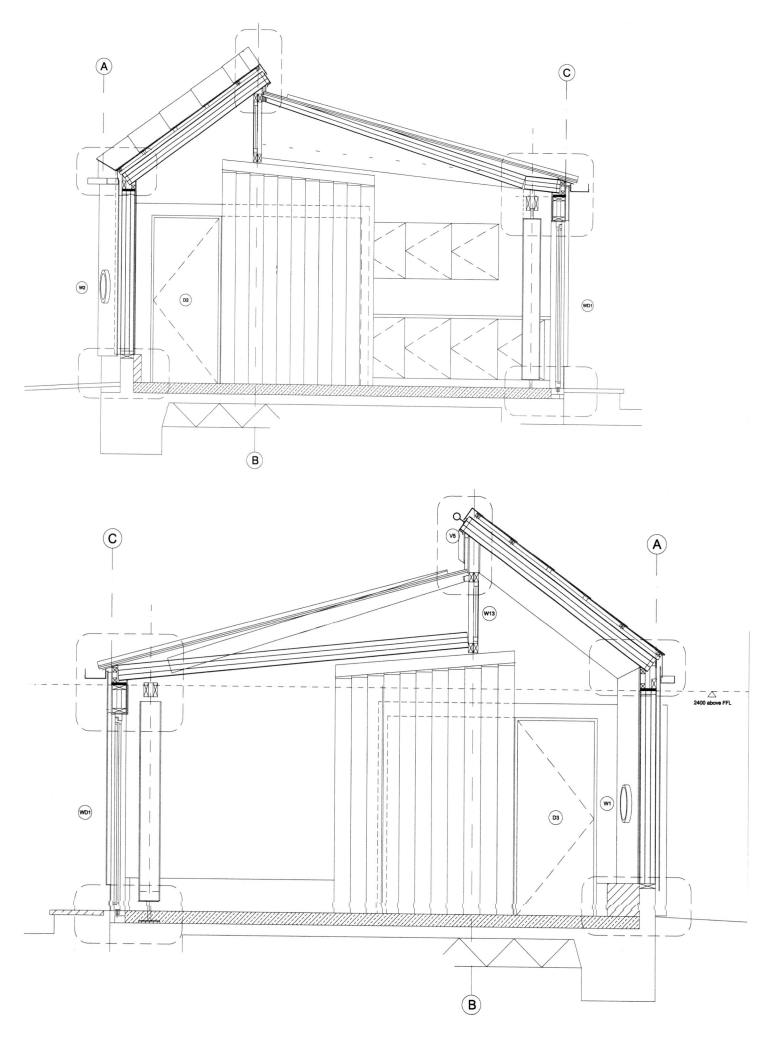

A   C

W2   D2   WD1

B

C   V5   A

W13

2400 above FFL

WD1   D3   W1

B

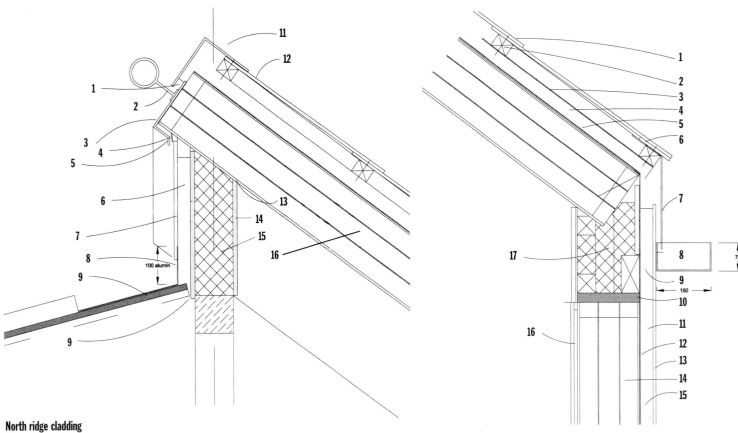

North ridge cladding

1. Ventilate/insect mesh
2. Stainless steel maintenance fixing hook
3. Aluminium flashing
4. Breather membrane
5. Ventilate
6. 38 x 50 softwood batten
7. Eternit board
8. Aluminium flashing
9. Polycarbonate
10. Ventalite timber truss
11. Aluminium flashing
12. Eternit tiles
13. Flexible joint
14. 4 mm polycoated cardboard
15. Cardboard insulation
16. Cardboard roof panels

South gutter detail

1. Eternit cladding tiles
2. Counter batten 38 x 50
3. Beather membrane
4. Long battens
5. Breather membrane
6. 38 x 50 softwood batten
7. Tilting fillet
8. Gutter
9. Ventilation through to cladding void
10. Wallplate
11. Vertical battens 38 x 50
12. Breather membrane
13. Eternit tiles
14. Cardboard panels
15. Ventilated void
16. 9 mm pinboard to 9 mm pinboard packing to cardboard structural panel
17. Cardboard insulation

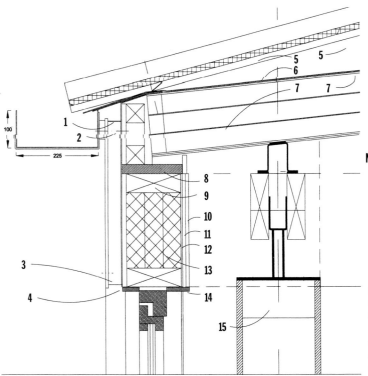

North gutter detail

1. Mesh closer
2. Fix gutter through battens
3. Klober mesh eaves closer
4. 12 mm thick hardwood bead
5. Ventilated void
6. Breather membrane
7. 166 mm composite cardboard
8. Wallplate
9. 150 x 50 softwood beam
10. 9 mm pinboard
11. 9 mm battens
12. 4 mm polycoated cardboard-pinned + glued to beam
13. Cardboard insulation
14. 12 mm deep ramin bead
15. Card column

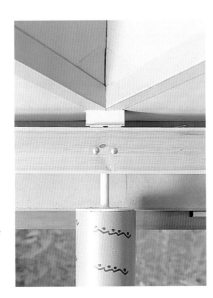

Bohlin Cywinski Jackson
Ledge House
The mountains of the rural Maryland, Maryland. USA

PHOTOGRAPHS: KARL A. BACKUS, AIA

Placed at the edge of a small plateau on a forested mountainside, the house overlooks a stream valley to the south. The manmade cut was the site of an earlier cabin. The clearing's upsweep edge is marked by stone ledges and a grove of pine trees.

By employing the logs, heavy timbers and stonework found in rustic buildings of the early 1990s and arranging the new structure along the south rim of the cut, a remarkably evocative forespace is created. This quarry-like place in the forest speaks of many activities from those of an entry court to a gathering space. On the other side of the extended log wall is a series of loosely arranged sheds that face the sun and overlook the valley, forest and stream below.

The shed roof structures are supported independently of the log walls and detailed with galvanized steel connectors. In the spirit of older camp structures, much of the framing for interior partitions and cabinets, as well as galvanized hardware and electrical fittings, have been exposed to view, adding the visual richness of the house.

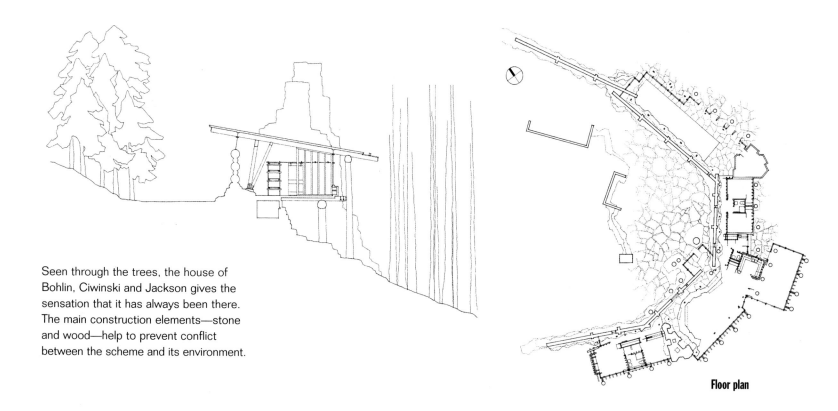

Seen through the trees, the house of Bohlin, Ciwinski and Jackson gives the sensation that it has always been there. The main construction elements—stone and wood—help to prevent conflict between the scheme and its environment.

Floor plan

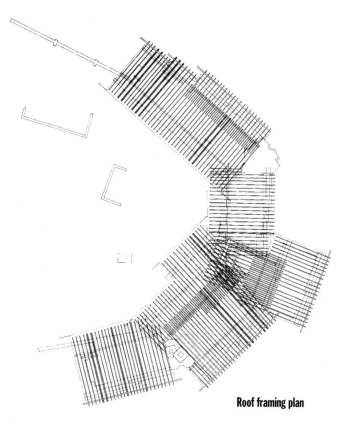

Roof framing plan

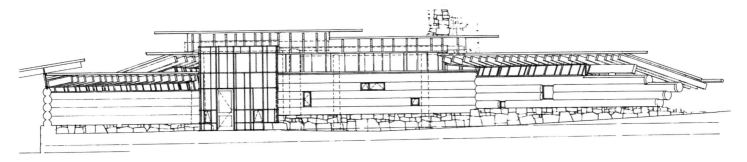

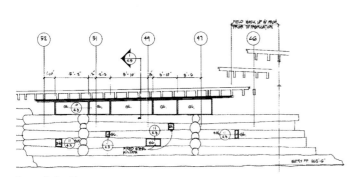

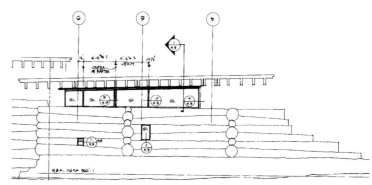

Log wall elevations

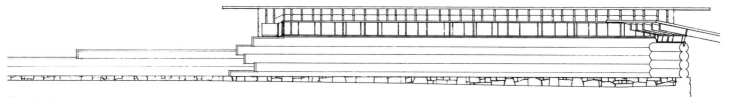

Pool elevation

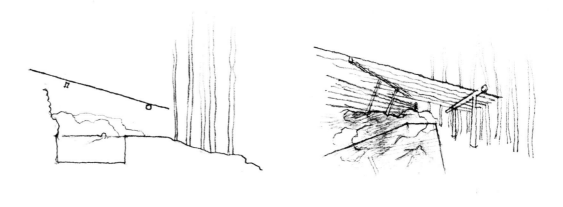

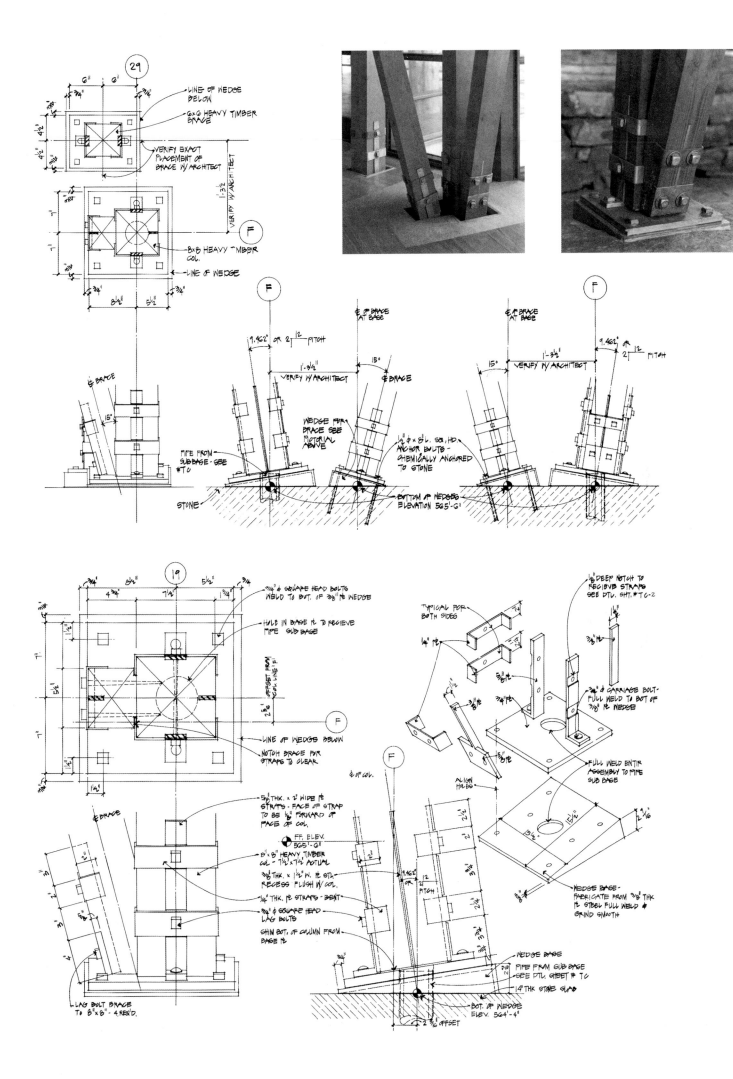

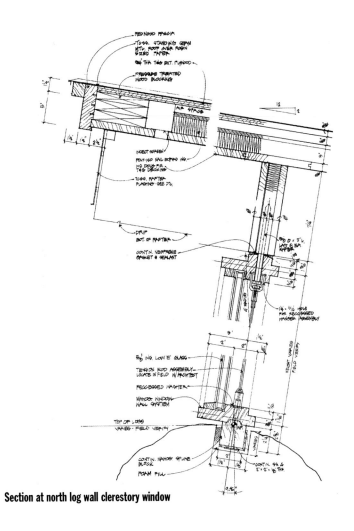

Section at north log wall clerestory window

169

Anne Lacaton & Jean Philippe Vassal

Maison à Lege

Cap Ferret, France

PHOTOGRAPHS: PHILIPPE RUAULT

This dwelling is in a wonderful location, giving directly onto Arcachon Bay and the Atlantic Ocean to the west of Bordeaux. It is a privileged space, one of the last undeveloped plots in an area with magnificent views over the water. Due to the characteristics of the land, an extension of sand dunes whose crest is 15 m above the level of the water, populated by fifty pines rising to 30 m, the architects made an effort to design a building that was able to adapt to and take advantage of its environment. The client brief consisted in building a dwelling on this site without damaging the environment and its qualities. The project maintains the density of the forest—no trees were cut down—and respects the natural undulation of the ground. Six trees go through the house. On the roof, transparent plane plastic sheets are tightened to the trunks with flexible rubber. They can glide on the edges of holes in the roof when the trees are moving with the wind.

The house is a platform of 210 sqm on a single level, built on pillars 2 to 4 metres high according to the slope of the land. This makes it possible to pass under the house and to have better views of the landscape from the interior of the dwelling. The structure is made of steel and the facades are of corrugated aluminium with openings of corrugated transparent fibreglass panels. The lower surface of the floor slab is covered with the same corrugated aluminium, which reflects the light reflected on the water. Finally, the front facade facing Arcachon Bay was left totally transparent, with sliding glass doors.

0 1   5   10

An aluminium spiral staircase leads to the 30 sqm terrace that is at the entrance to this unusual dwelling. In addition to creating an unusual visual effect, the pines incorporated into the building help to camouflage it in the site.

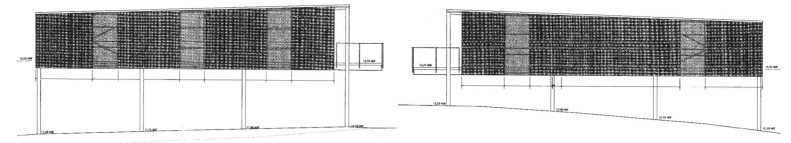

West elevation   **East elevation**

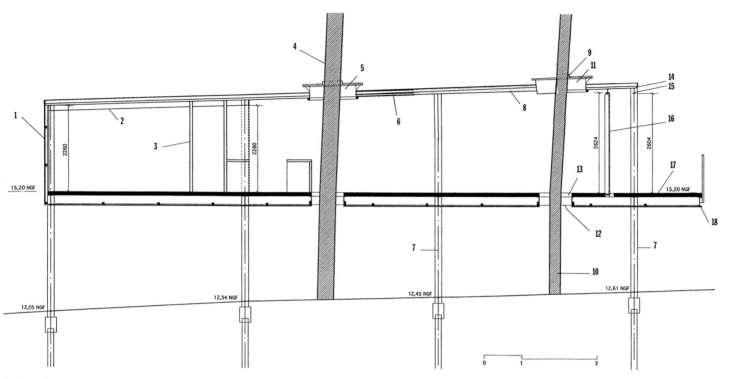

South-north longitudinal section

1. Aluminium shape
2. Plasterboard ceiling
3. Partition with plasterboard
4. Pine trunk 20
5. Outer ribbing
6. Corrugated aluminium shape
7. Hea 180
8. Hea 100
9. Steel hoop around the trunk + strip of flexible rubber
10. Pine trunk 27

11. Outer ribbing
12. Polyethylene fabric
13. Plywood plaque
14. Drainage strip
15. Galvanised sheet folded into a L-shape to protect door and window frames
16. Sliding windows on three aluminium rails
17. Concrete floor
18. Galvanised steel piece to avoid water entering

Although between the volume of the house and the ground level the trees that go through the house can be confused with the pillars, in the interior the latter play an important decorative role.

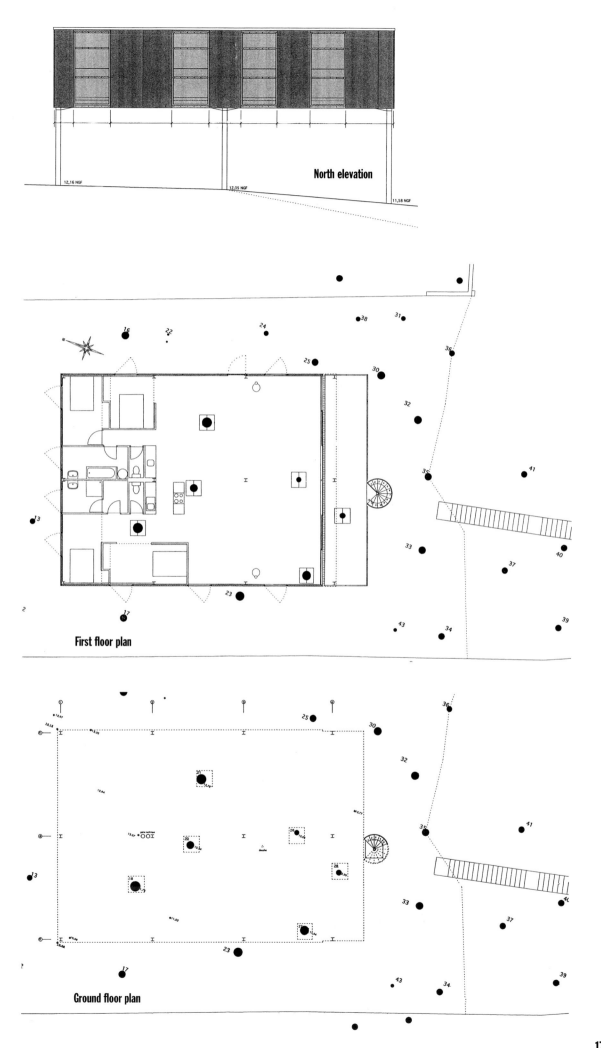

North elevation

12,16 NGF

12,05 NGF

11,58 NGF

First floor plan

Ground floor plan

Nader Khalili
Hesperia Museum and Nature Center
Hesperia, California. USA

On April 19th, Earth Day weekend 1996, the City of Hesperia Recreation and Park District broke ground to construct the Hesperia Museum and Nature Center, the first of it's kind in the world, legally permitted to be built totally of earthen construction, as part of the Desert Village, Rodeo/Arena and other futuristic programs in Hesperia.

The complex of fourteen domes and two vaults in earth and ceramics is designed by internationally renowned architect and author Nader Khalili, founder of the California Institute of Earth Art and Architecture (Cal-Earth), who has devoted the last 20 years to developing simple disaster safe Earth Architecture technologies along with his associates and apprentices.

The project was approved by the Hesperia Building Department in consultation with the ICBO (International Conference of Building Officials) after successful static and dynamic load testing for wind, snow and earthquakes on full scale prototypes at Cal-Earth in Hesperia. The extensive engineering analysis and design, and the tests, were devised by Phill Vittore, a world specialist in dome engineering and structural design.

The architect's innovations of Superadobe (mile-long sandbags and barbed wire construction) and Ceramic Houses (fired in-place adobe structu-res) are seen as solutions to global deforestation by ecological organizations and foundations who have lent their sponsorship along with other organisations to work at Cal-Earth.

The United Nations, with whom Khalili has built housing prototypes using these technologies, see these as the answer to affordable housing worldwide and emergency relief housing because of their efficient simplicity.

The Hesperia Museum and Nature Center sets the most important precedent for building with earth in the American fast-track mainstream in a harsh climate. The project demonstrates how, with the simplest of elements, Earth, Wind, Water, and Fire, and with the ancient earth architecture techniques developed in the deserts of Iran and the Middle East, and traditional cultures from the Native Americans of the far west to China in the far east, whole communities could be developed in total harmony with the environment.

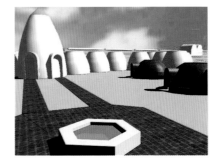

PHOTOGRAPHS: CAL EARTH

176

The Superadobe (sandbag and barbed wire construction system) has been developed to create an instant rammed-earth type wall.

Standard or long tubular bags are pumped or hand-filled with on-site earth (unprocessed or stabilized) and coiled in place to from walls, structural arches, domes and vaults, reinforced with strands of barbed wire.